MATHEMATICS RESEARCH DEVELOPMENTS

RLS WIENER SMOOTHER FROM RANDOMLY DELAYED OBSERVATIONS IN LINEAR DISCRETE-TIME SYSTEMS

MATHEMATICS RESEARCH DEVELOPMENTS

Additional books in this series can be found on Nova's website under the Series tab.

Additional e-books in this series can be found on Nova's website under the e-book tab.

MATHEMATICS RESEARCH DEVELOPMENTS

RLS WIENER SMOOTHER FROM RANDOMLY DELAYED OBSERVATIONS IN LINEAR DISCRETE-TIME SYSTEMS

SEIICHI NAKAMORI

nova publishers
New York

For permission to use material from this book please contact us:
Telephone 631-231-7269; Fax 631-231-8175
Web Site: http://www.novapublishers.com

NOTICE TO THE READER

Additional color graphics may be available in the e-book version of this book.

Library of Congress Cataloging-in-Publication Data

ISBN: 978-1-62417-818-4

Published by Nova Science Publishers, Inc. † New York

CONTENTS

Acknowledgements

This monograph consists of my original research, which has not published elsewhere. Upon this publication, I would like to express my gratitude to NOVA Science publishers.

Summary

This monograph designs the new recursive least-squares (RLS) Wiener filter and fixed-point smoother from randomly delayed observed values by one sampling time in linear discrete-time stochastic systems. The observation equation is given by $y(k) = (1-\gamma(k))\bar{y}(k) + \gamma(k)\bar{y}(k-1)$(see (2.1)). The actual observed value consists of the observed value $\bar{y}(k-1)$ with the probability $p(k)$ and $\bar{y}(k)$ with the probability $1-p(k)$. $\bar{y}(k)$ is given as a sum of the signal $z(k)$ and the white observation noise. The estimators require the information of the system matrix Φ concerned with the state vector$x(k)$, the observation matrix H, the variance $K(k,k)$ of the state vector $x(k)$, the variance $R(k)$ of the observation noise, the probability $p(k) = P\{\gamma(k) = 1\}$ that the signal exists in the uncertain observation equation and the $(2,2)$ element $[P(k|j)]_{2,2}$ of the conditional probability of $\gamma(k)$, given $\gamma(j)$.

Keywords: Synthesis of stochastic systems, RLS Wiener fixed-point smoother, randomly delayed observations, conditional probability, covariance information

Chapter 1

Introduction

The estimation problem, given the uncertain observations, is important in the detection and estimation for communication systems. The recursive least-squares (RLS) estimation technique is proposed for the state variable in dynamical systems with the uncertain observations [1].[2]. Here, the uncertain observations mean that some observations do not contain the signal and consist of observation noise only. The probability for the existence of the signal in the observations is available.

There are investigations, using different approaches, on the state estimation problem for system models with randomly varying time-delays ([3]-[6]).

In [7], the RLS Wiener filter and fixed-point smoother are designed in linear discrete-time systems. The estimators require the information of the system matrix, the observation vector and the variance of the state vector in the signal generating model, and the variance of white Gaussian observation noise. By appropriate choices of the observation vector and the state variables, the state-space model corresponding to the autoregressive moving average (ARMA) model is introduced. Here, some elements of the system matrix consist of the AR parameters.

In [8], using the covariance information, the estimation method for the signal with the uncertain observations is considered in linear discrete-time stochastic systems. The estimation technique using the covariance information has been presented in the context or the detection and estimation of the signal [9]. The recursive Wiener filter with the uncertain observations is devised in linear stochastic systems [8].

In [10], the least-squares linear estimation problem of signals from randomly delayed observations is considered when the additive white noise is correlated with the signal. The delay values are treated as unknown variables, modeled by a binary white noise with values zero or one; these values indicate that the measurements arrive in time or they are delayed by one sampling time. A recursive one-stage prediction and filtering algorithms are obtained by an innovation approach and do not use the state-space model of the signal. It is assumed that both, the auto-covariance functions of the signal and the cross-covariance function between the signal and the observation noise are expressed in a semi-degenerate kernel form; using this information and the delay probabilities, the estimates are calculated recursively.

In [11], the least-squares linear filtering, fixed-point smoothing and fixed-interval smoothing algorithms of discrete-time signals are obtained from observations, perturbed by additive white noise, which can be randomly delayed by one sampling time. It is assumed that the Bernoulli random variables characterizing delay measurements are correlated in consecutive time instants. Using an innovation approach, without requiring the state-space model generating the signa1, the estimation algorithms probabilities and the correlation function of the Bernoulli variables. In [12], [13], the dynamic Markov jump filter is proposed for the observations with random jump delays of the state vector. In [14], the RLS Wiener fixed-point smoother and filter are proposed in the estimation problem from randomly delayed observations by one sampling time.

Nakamori et al. [15], based on the innovation approach, proposed the RLS Wiener fixed-point smoother and filter in linear discrete-time stochastic systems. Here, the observation equation is given by $y(k) = \gamma(k)z(k) + v(k)$, $z(k) = Hx(k)$, where $\{\gamma(k)\}$ is a binary switching sequence with conditional probability and a Markovian uncertainty. Here, $\{\gamma(k)\}$ is a mixture sequences that model the transmission of data in multichannels. The innovation process is given by $\upsilon(s) = y(s) - \hat{y}(s, s-1), \hat{y}(s, s-1) = P_{2,2}(s)H\Phi\hat{x}(s-1, s-1)$ in terms of the$(2,2)$ element $[P(k|j)]_{2,2}$ [16] of the conditional probability of $\gamma(k)$, given $\gamma(j)$. Similarly, in [17], the RLS Wiener l-step ahead predictor and filter are proposed for the observation equation $y(k) = \gamma(k)z(k) + v(k)$ by using the same information as in [15].

This monograph designs the new recursive least-squares (RLS) Wiener filter and fixed-point smoother from randomly delayed observed values by one sampling time in linear discrete-time stochastic systems. The observation equation is given by $y(k) = (1-\gamma(k))\bar{y}(k) + \gamma(k)\bar{y}(k-1)$(see (2.1)). The actual observed value consists of the observed value $\bar{y}(k-1)$ with the probability $p(k)$ and $\bar{y}(k)$ with the probability $1-p(k)$. $\bar{y}(k)$ is given as a sum of the signal $z(k)$ and the white observation noise. The estimators require the information of the system matrix Φ concerned with the state vector$x(k)$, the observation matrix H, the variance $K(k,k)$ of the state vector $x(k)$, the variance $R(k)$ of the observation noise, the probability $p(k) = P\{\gamma(k) = 1\}$ that the signal exists in the uncertain observation equation and the $(2,2)$ element $[P(k|j)]_{2,2}$ of the conditional probability of $\gamma(k)$, given $\gamma(j)$. Also, the filtering error variance functions of the signal and white observation noise are formulated.

A numerical simulation example, in chapter 5, shows the estimation characteristics of the current filter and the fixed-point smoother with the randomly delayed observed values by one sampling time.

Chapter 2

Least-squares Fixed-point Smoothing Problem

Let an observation equation be given by

$$\begin{aligned} y(k) &= (1-\gamma(k))\bar{y}(k) + \gamma(k)\bar{y}(k-1), \quad E[\gamma(k)] = p(k), \\ \bar{y}(k) &= z(k) + v(k), \quad z(k) = Hx(k) \end{aligned} \tag{2.1}$$

where $z(k)$ is a signal, $x(k)$ is the $n \times 1$ zero-mean state vector and H is the $m \times n$ observation matrix. It is assumed that the observation at each time $k > 1$ can be either delayed by one sampling period, with a known probability $p(k)$, or updated, with probability $1 - p(k)$. Usually, in applications of communication networks, $\{\ \gamma(k); \quad k > 1\ \}$ represents the random delay from sensor to controller and the assumption of one-step sensor delay is based on the reasonable supposition that the induced data latency from the sensor to the controller is restricted not to exceed the sampling period [5].

- The sequence $\{v(k)\}$ is white noise with zero-mean and the variance of $v(k)$ is $R(k)$, that is,

$$E[v(k)v^T(s)] = R(k)\delta_K(k-s), \tag{2.2}$$

where $\delta_K(\cdot)$ denotes the Kronecker delta function.

- The random sequence $\{\gamma(k)\}$, which describes the uncertainty in the observations, has the following stochastic properties [16]:

(P-1) $\gamma(k)$ is a discrete-time random variable which takes the values 0 or 1 with $P\{\gamma(k) = 1\} = p(k)$. Therefore, $p(k)$ represents the probability that the observed value $y(k)$ contains the signal$z(k)$; this probability is assumed to be nonzero.

(P-2) The noise $\{\gamma(k)\}$ is a sequence of random variables with initial probability vector $(1 - p(0), p(0))^T$ and conditional probability matrix $P(k|j)$. The $(2,2)$ element of the conditional probability matrix of $\gamma(k)$ given$\gamma(j)$, is independent of j, for $j < k$, that is

$$[P(k|j)]_{2,2} = \frac{E[\gamma(j)\gamma(k)]}{P\{\gamma(j) = 1\}} = p_{2,2}(k), j = 0, \cdots, k-1. \tag{2.3}$$

- The state process $\{x(k)\}$ and the sequences $\{\gamma(k)\}$ and $\{v(k)\}$ are mutually independent.

By denoting

$$\begin{aligned} \bar{\gamma}(k) &= \left[\begin{array}{cc} \gamma(k)I_{m\times m} & (1-\gamma(k))I_{m\times m} \end{array} \right], \quad \bar{z}(k) = \left[\begin{array}{c} z(k-1) \\ z(k) \end{array} \right], \\ \bar{v}(k) &= \bar{\gamma}(k)V(k), \quad V(k) = \left[\begin{array}{c} v(k-1) \\ v(k) \end{array} \right]. \end{aligned} \tag{2.4}$$

from (2.1), we obtain

$$y(k) = \bar{\gamma}(k)\bar{z}(k) + \bar{v}(k). \tag{2.5}$$

By denoting

$$\bar{H} = \left[\begin{array}{cc} H & 0 \\ 0 & H \end{array} \right], \quad \bar{x}(k) = \left[\begin{array}{c} x(k-1) \\ x(k) \end{array} \right], \tag{2.6}$$

from (2.1) and (2.4), the observation equation (2.5) is also written as

$$y(k) = \bar{\gamma}(k)\bar{H}\bar{x}(k) + \bar{v}(k), \tag{2.7}$$

since

$$\bar{z}(k) = \left[\begin{array}{c} z(k-1) \\ z(k) \end{array} \right] = \left[\begin{array}{c} Hx(k-1) \\ Hx(k) \end{array} \right] = \bar{H}\bar{x}(k). \tag{2.8}$$

Let $K_x(k,s) = K_x(k-s)$ represent the auto-covariance function of the state vector $x(k)$ in wide-sense stationary stochastic systems [14], and let $K_x(k,s)$ be expressed in the form of

$$K_x(k,s) = \begin{cases} A(k)B^T(s), & 0 \le s \le k, \\ B(s)A^T(k), & 0 \le k \le s, \end{cases} \tag{2.9}$$

$A(k) = \Phi^k, \quad B^T(s) = \Phi^{-s}K_x(s,s)$. Here, Φ is the transition matrix of $x(k)$.

Let the state-space model for $x(k)$ be described as

$$x(k+1) = \Phi x(k) + Bw(k), \quad E[w(k)w^T(s)] = Q\delta_K(k-s), \tag{2.10}$$

where B is n by l input matrix and $w(k)$ is white input noise with the auto-covariance function of (2.9). Let $\bar{\Phi}$ represent the system matrix for $\bar{x}(k)$. From

$$\begin{bmatrix} x(k) \\ x(k+1) \end{bmatrix} = \begin{bmatrix} 0 & I \\ 0 & \Phi \end{bmatrix} \begin{bmatrix} x(k-1) \\ x(k) \end{bmatrix} + \begin{bmatrix} 0 \\ B \end{bmatrix} \begin{bmatrix} 0 \\ w(k) \end{bmatrix}, \tag{2.11}$$

$\bar{\Phi}$ is given by

$$\bar{\Phi} = \begin{bmatrix} 0 & I \\ 0 & \Phi \end{bmatrix}. \tag{2.12}$$

Let $K_{\bar{x}}(k,s)$ represent the auto-covariance function indexauto-covariance function of $\bar{x}(k)$. $K_{\bar{x}}(k,s)$ is expressed in the form of

$$K_{\bar{x}}(t,s) = \begin{cases} \bar{A}(k)\bar{B}^T(s), & 0 \le s \le k, \\ \bar{B}(s)\bar{A}^T(t), & 0 \le k \le s, \end{cases} \tag{2.13}$$

$\bar{A}(k) = \bar{\Phi}^k, \quad \bar{B}^T(s) = \bar{\Phi}^{-s}K_{\bar{x}}(s,s)$. Here, $K_{\bar{x}}(s,s) = K_{\bar{x}}(0)$ is expressed as

$$K_{\bar{x}}(s,s) = E\left[\begin{bmatrix} x(s-1) \\ x(s) \end{bmatrix} \begin{bmatrix} x^T(s-1) & x^T(s) \end{bmatrix}\right] = \begin{bmatrix} K_x(0) & K_x(0)\Phi^T \\ \Phi K_x(0) & K_x(0) \end{bmatrix}. \tag{2.14}$$

Let $K_V(\cdot,\cdot)$ denote the auto-covariance function of $V(k)$. From (2.2) and (2.4), $K_V(k,k)$ and $K_V(k+1,k)$ are given by

$$K_V(k,k) = E\left[V(k)V^T(k)\right] = E\left[\begin{bmatrix} v(k-1) \\ v(k) \end{bmatrix} \begin{bmatrix} v^T(k-1) & v^T(k) \end{bmatrix}\right] = \begin{bmatrix} R & 0 \\ 0 & R \end{bmatrix},$$

$$K_V(k+1,k) = E\left[V(k+1)V^T(k)\right] = E\left[\begin{bmatrix} v(k) \\ v(k+1) \end{bmatrix} \begin{bmatrix} v^T(k-1) & v^T(k) \end{bmatrix}\right] = \begin{bmatrix} 0 & R \\ 0 & 0 \end{bmatrix}. \tag{2.15}$$

From (2.15) and $K_V(k+i,k)=0$, $i\geq 2$, the auto-covariance function $K_V(k,s)$ is given by

$$K_V(k,s)=\begin{cases} A_V(k)B_V^T(s), & 0\leq s\leq k,\\ B_V(s)A_V^T(t), & 0\leq k\leq s,\end{cases}$$
$$A_V(k)=\Phi_V^k,\quad B^T(s)=\Phi_V^{-s}K_V(s,s),\quad \Phi_V=\begin{bmatrix} 0 & I\\ 0 & 0\end{bmatrix}. \tag{2.16}$$

Let the fixed-point smoothing estimate $\hat{\bar{x}}(k,L)$ of $\bar{x}(k)$ at the fixed point k be expressed by

$$\hat{\bar{x}}(k,L)=\sum_{i=1}^{L}h(k,i,L)y(i) \tag{2.17}$$

in terms of the observed values $\{\ y(i),\ \ 1\leq i\leq L\ \}$. In (2.17), $h(k,i,L)$ is a time-varying impulse response function.

Let us consider the estimation problem, which minimizes the mean-square value (MSV)

$$J=E[||\bar{x}(k)-\hat{\bar{x}}(k,L)||^2] \tag{2.18}$$

of the fixed-point smoothing error. From an orthogonal projection lemma [18],

$$\bar{x}(k)-\sum_{i=1}^{L}h(k,i,L)y(i)\perp y(s),\quad 1\leq s\leq L, \tag{2.19}$$

the impulse response function satisfies the Wiener-Hopf equation

$$E[\bar{x}(k)y^T(s)]=\sum_{i=1}^{L}h(k,i,L)E[y(i)y^T(s)]. \tag{2.20}$$

Here '$\perp$' denotes the notation of the orthogonality.

By putting $\bar{p}(k)=\left[\ p(k)I_{m\times m}\quad (1-p(k))I_{m\times m}\ \right]$, from (2.4) and (2.5), the left hand side of (2.20) is rewritten as

$$\begin{aligned} &E[\bar{x}(k)y^T(s)]=E[\bar{x}(k)(\bar{\gamma}(s)\bar{z}(s)+\bar{v}(s))^T]\\ &=E[\bar{x}(k)\bar{z}^T(s)\bar{\gamma}^T(s)]\\ &=E[\bar{x}(k)\bar{x}^T(s)]\bar{H}^T[\ p(s)I_{m\times m}\quad (1-p(s))\ I_{m\times m}]^T\\ &=K_{\bar{x}}(k,s)\bar{H}^T\bar{p}^T(s).\end{aligned} \tag{2.21}$$

Also, from (2.4) and (2.7), $E[y(i)y^T(s)]$ is rewritten as

$$\begin{aligned}E[y(i)y^T(s)] &= E[(\bar{\gamma}(i)\bar{H}\bar{x}(i)+\bar{v}(i))(\bar{\gamma}(s)\bar{H}\bar{x}(s)+\bar{v}(s))^T] \\ &= E_\gamma[\bar{\gamma}(i)(\bar{H}K_{\bar{x}}(i,s)\bar{H}^T+K_V(i,s))\bar{\gamma}^T(s)].\end{aligned} \tag{2.22}$$

Here, $E_\gamma[\cdot]$ denotes the notation of expectation with respect to $\gamma(\cdot)$. Substituting (2.21) and (2.22) into (2.20), we obtain

$$K_{\bar{x}}(k,s)\bar{H}^T\bar{p}^T(s) = \sum_{i=1}^{L} h(k,i,L)E_\gamma[\bar{\gamma}(i)(\bar{H}K_{\bar{x}}(i,s)\bar{H}^T+K_V(i,s))\bar{\gamma}^T(s)]. \tag{2.23}$$

Let $V(k)$ satisfy the state equation

$$V(k+1) = \Phi_V V(k)+U(k), \quad E[U(k)U^T(s)] = R_U(k)\delta_K(k-s). \tag{2.24}$$

It is clear that the following relation holds.

$$K_V(i,s) = \Phi_V K_V(i,s)\Phi_V^T+\Phi_V K_{VU}(i,s)+K_{UV}(i,s)\Phi_V^T+R_U\delta_K(i-s). \tag{2.25}$$

Substituting (2.25) into (2.23), we obtain the equation for the impulse response function to satisfy

$$\begin{aligned}&h(k,s,L)\bar{R}_U(s) = K_{\bar{x}}(k,s)\bar{H}^T\bar{p}^T(s) \\ &-\sum_{i=1}^{L} h(k,i,L)E_\gamma[\bar{\gamma}(i)(\bar{H}K_{\bar{x}}(i,s)\bar{H}^T+\Phi_V K_V(i,s)\Phi_V^T+\Phi_V K_{VU}(i,s) \\ &+K_{UV}(i,s)\Phi_V^T)\bar{\gamma}^T(s)], \\ &\bar{R}_U(s) = E_\gamma[\bar{\gamma}(s)R_U(s)\bar{\gamma}^T(s)].\end{aligned} \tag{2.26}$$

Chapter 3

RLS Wiener Estimation Algorithms

Under the linear least-squares estimation problem of the signal $z(k)$ in chapter 2, Theorem 1 presents the RLS Wiener fixed-point smoothing and filtering algorithms for the signal $z(k)$ and the state vector$x(k)$. These algorithms are derived, starting with (2.26), by the iterative use of the invariant imbedding method.

Theorem 1. Let us consider the randomly delayed observation equation described by (2.1). Let the probability $p(k)$ and the $(2,2)$ element $P_{2,2}(k)$ of the conditional probability matrix $P(k|j)$ be given. Let the auto-covariance function $K_x(k,s)$ of the state vector $x(k)$ be expressed by (2.9), and let the variance of white observation noise be R. Then, the RLS Wiener algorithms for the fixed-point smoothing at the fixed point k and filtering estimate of the signal $z(k)$ consist of (3.1)-(3.134) in linear discrete-time stochastic systems with randomly delayed observations.

Fixed-point smoothing estimate of the signal $z(k)$: $\hat{z}(k,L) = H\hat{x}(k,L)$

$$\hat{x}(k,L) = \begin{bmatrix} 0 & I \end{bmatrix} \hat{\hat{x}}(k,L), \quad \hat{\hat{x}}(k,L) = \begin{bmatrix} \hat{x}(k-1,L) \\ \hat{x}(k,L) \end{bmatrix} \tag{3.1}$$

Fixed-point smoothing estimate of $\bar{x}(k)$: $\hat{\bar{x}}(k,L)$

$$\begin{aligned}
&\hat{\bar{x}}(k,L)=\hat{\bar{x}}(k,L-1)+h(k,L,L)(y(L)-\bar{p}_{2,2}(L)(\bar{H}\bar{\Phi}\bar{O}_{11}(L-1)+\Phi_V^2\bar{O}_{12}(L-1)\\
&+\Phi_V\bar{O}_{13}(L-1))-\bar{p}(L)(\bar{H}\bar{\Phi}\bar{O}_{21}(L-1)+\Phi_V^2\bar{O}_{22}(L-1)+\Phi_V\bar{O}_{23}(L-1))\\
&+\bar{p}_{2,2}(L)(\bar{H}\bar{\Phi}\bar{O}_{31}(L-1)+\Phi_V^2\bar{O}_{32}(L-1)+\Phi_V\bar{O}_{33}(L-1))),\\
&\bar{H}=\begin{bmatrix} H & 0\\ 0 & H\end{bmatrix},\quad \bar{\Phi}=\begin{bmatrix} 0 & I\\ 0 & \Phi\end{bmatrix},\quad \Phi_V=\begin{bmatrix} 0 & I\\ 0 & 0\end{bmatrix},\\
&\bar{p}(L)=\begin{bmatrix} p(L)I_m & (1-p(L))I_m\end{bmatrix},\quad \bar{p}_{2,2}(L)=\begin{bmatrix} p_{2,2}(L)I_m & (1-p_{2,2}(L))I_m\end{bmatrix},
\end{aligned} \tag{3.2}$$

Smoother gain for $\hat{\bar{x}}(k,L)$: $h(k,L,L)$

$$\begin{aligned}
&h(k,L,L)=[K_{\bar{x}}(k,k)(\bar{\Phi}^T)^{L-k}\bar{H}^T\bar{p}^T(L)-(\bar{C}_{1,1}(k,L-1)\bar{\Phi}^T\bar{H}^T+\bar{C}_{12}(k,L-1)(\Phi_V^T)^2\\
&+\bar{C}_{13}(k,L-1)\Phi_V^T)\bar{p}_{2,2}^T(L)-(\bar{C}_{2,1}(k,L-1)\bar{\Phi}^T\bar{H}^T+\bar{C}_{22}(k,L-1)(\Phi_V^T)^2\\
&+\bar{C}_{23}(k,L-1)\Phi_V^T)\bar{p}^T(L)+(\bar{C}_{3,1}(k,L-1)\bar{\Phi}^T\bar{H}^T+\bar{C}_{32}(k,L-1)(\Phi_V^T)^2\\
&+\bar{C}_{33}(k,L-1)\Phi_V^T)\bar{p}_{2,2}^T(L))]\Gamma^{-1}(L).
\end{aligned} \tag{3.3}$$

$$\begin{aligned}
&\bar{C}_{11}(k,L)=\bar{C}_{11}(k,L-1)\bar{\Phi}^T+h(k,L,L)[\,p(L)I_m\quad 0\,]\bar{H}K_{\bar{x}}(L,L)\\
&-h(k,L,L)\{\bar{p}_{2,2}(L)[\bar{H}\bar{\Phi}S_{1111}(L-1)\bar{\Phi}^T+\Phi_V^2S_{1121}(L-1)\bar{\Phi}^T+\Phi_VS_{1131}(L-1)\bar{\Phi}^T]\\
&+\bar{p}(L)[\bar{H}\bar{\Phi}S_{2111}(L-1)\bar{\Phi}^T+\Phi_V^2S_{2121}(L-1)\bar{\Phi}^T+\Phi_VS_{2131}(L-1)\bar{\Phi}^T]\\
&-\bar{p}_{2,2}(L)[\bar{H}\bar{\Phi}S_{3111}(L-1)\bar{\Phi}^T+\Phi_V^2S_{3121}(L-1)\bar{\Phi}^T+\Phi_VS_{3131}(L-1)\bar{\Phi}^T]\}
\end{aligned} \tag{3.4}$$

$$\begin{aligned}
&\bar{C}_{12}(k,L)=\bar{C}_{12}(k,L-1)\Phi_V^T+h(k,L,L)[\,p(L)I_m\quad 0\,]\Phi_VK_V(L,L)\\
&-h(k,L,L)\{\bar{p}_{2,2}(L)[\bar{H}\bar{\Phi}S_{1112}(L-1)\Phi_V^T+\Phi_V^2S_{1122}(L-1)\Phi_V^T+\Phi_VS_{1132}(L-1)\Phi_V^T]\\
&+\bar{p}(L)[\bar{H}\bar{\Phi}S_{2112}(L-1)\Phi_V^T+\Phi_V^2S_{2122}(L-1)\Phi_V^T+\Phi_VS_{2132}(L-1)\Phi_V^T]\\
&-\bar{p}_{2,2}(L)[\bar{H}\bar{\Phi}S_{3112}(L-1)\Phi_V^T+\Phi_V^2S_{3122}(L-1)\Phi_V^T+\Phi_VS_{3132}(L-1)\Phi_V^T]\}
\end{aligned} \tag{3.5}$$

$$\begin{aligned}
&\bar{C}_{13}(k,L)=\bar{C}_{13}(k,L-1)\Phi_V^T+h(k,L,L)[\,p(L)I_m\quad 0\,]R_U(L)\\
&-h(k,L,L)\{\bar{p}_{2,2}(L)[\bar{H}\bar{\Phi}S_{1113}(L-1)\Phi_V^T+\Phi_V^2S_{1123}(L-1)\Phi_V^T+\Phi_VS_{1133}(L-1)\Phi_V^T]\\
&+\bar{p}(L)[\bar{H}\bar{\Phi}S_{2113}(L-1)\Phi_V^T+\Phi_V^2S_{2123}(L-1)\Phi_V^T+\Phi_VS_{2133}(L-1)\Phi_V^T]\\
&-\bar{p}_{2,2}(L)[\bar{H}\bar{\Phi}S_{3113}(L-1)\Phi_V^T+\Phi_V^2S_{3123}(L-1)\Phi_V^T+\Phi_VS_{3133}(L-1)\Phi_V^T]\}
\end{aligned} \tag{3.6}$$

$$\begin{aligned}
&\bar{C}_{21}(k,L)=\bar{C}_{21}(k,L-1)\bar{\Phi}^T+h(k,L,L)[\,0\quad I_m\,]\bar{H}K_{\bar{x}}(L,L)\\
&-h(k,L,L)\{\bar{p}_{2,2}(L)[\bar{H}\bar{\Phi}S_{1211}(L-1)\bar{\Phi}^T+\Phi_V^2S_{1221}(L-1)\bar{\Phi}^T+\Phi_VS_{1231}(L-1)\bar{\Phi}^T]\\
&+\bar{p}(L)[\bar{H}\bar{\Phi}S_{2211}(L-1)\bar{\Phi}^T+\Phi_V^2S_{2221}(L-1)\bar{\Phi}^T+\Phi_VS_{2231}(L-1)\bar{\Phi}^T]\\
&-\bar{p}(L)[\bar{H}\bar{\Phi}S_{3211}(L-1)\bar{\Phi}^T+\Phi_V^2S_{3221}(L-1)\bar{\Phi}^T+\Phi_VS_{3231}(L-1)\bar{\Phi}^T]\}
\end{aligned} \tag{3.7}$$

$$\begin{aligned}
&\bar{C}_{22}(k,L)=\bar{C}_{22}(k,L-1)\Phi_V^T+h(k,L,L)[\,0\quad I_m\,]\Phi_VK_V(L,L)\\
&-h(k,L,L)\{\bar{p}_{2,2}(L)[\bar{H}\bar{\Phi}S_{1212}(L-1)\Phi_V^T+\Phi_V^2S_{1222}(L-1)\Phi_V^T+\Phi_VS_{1232}(L-1)\Phi_V^T]\\
&+\bar{p}(L)[\bar{H}\bar{\Phi}S_{2212}(L-1)\Phi_V^T+\Phi_V^2S_{2222}(L-1)\Phi_V^T+\Phi_VS_{2232}(L-1)\Phi_V^T]\\
&-\bar{p}_{2,2}(L)[\bar{H}\bar{\Phi}S_{3212}(L-1)\Phi_V^T+\Phi_V^2S_{3222}(L-1)\Phi_V^T+\Phi_VS_{3232}(L-1)\Phi_V^T]\}
\end{aligned} \tag{3.8}$$

$$\bar{C}_{23}(k,L)=\bar{C}_{23}(k,L-1)\Phi_V^T+h(k,L,L)[\ 0 \quad I_m\ ]R_U(L)$$
$$-h(k,L,L)\{\bar{p}_{2,2}(L)[\bar{H}\bar{\Phi}S_{1213}(L-1)\Phi_V^T+\Phi_V^2S_{1223}(L-1)\Phi_V^T+\Phi_VS_{1233}(L-1)\Phi_V^T]$$
$$+\bar{p}(L)[\bar{H}\bar{\Phi}S_{2213}(L-1)\Phi_V^T+\Phi_V^2S_{2223}(L-1)\Phi_V^T+\Phi_VS_{2233}(L-1)\Phi_V^T]$$
$$-\bar{p}_{2,2}(L)[\bar{H}\bar{\Phi}S_{3213}(L-1)\Phi_V^T+\Phi_V^2S_{3223}(L-1)\Phi_V^T+\Phi_VS_{3233}(L-1)\Phi_V^T]\} \tag{3.9}$$

$$\bar{C}_{31}(k,L)=\bar{C}_{31}(k,L-1)\bar{\Phi}^T+h(k,L,L)[\ 0 \quad p(L)I_m\ ]\bar{H}K_{\bar{x}}(L,L)$$
$$-h(k,L,L)\{\bar{p}_{2,2}(L)[\bar{H}\bar{\Phi}S_{1311}(L-1)\bar{\Phi}^T+\Phi_V^2S_{1321}(L-1)\bar{\Phi}^T+\Phi_VS_{1331}(L-1)\bar{\Phi}^T]$$
$$+\bar{p}(L)[\bar{H}\bar{\Phi}S_{2311}(L-1)\bar{\Phi}^T+\Phi_V^2S_{2321}(L-1)\bar{\Phi}^T+\Phi_VS_{2331}(L-1)\bar{\Phi}^T]$$
$$-\bar{p}_{2,2}(L)[\bar{H}\bar{\Phi}S_{3311}(L-1)\bar{\Phi}^T+\Phi_V^2S_{3321}(L-1)\bar{\Phi}^T+\Phi_VS_{3331}(L-1)\bar{\Phi}^T]\} \tag{3.10}$$

$$\bar{C}_{32}(k,L)=\bar{C}_{32}(k,L-1)\Phi_V^T+h(k,L,L)[\ 0 \quad p(L)I_m\ ]\Phi_VK_V(L,L)$$
$$-h(k,L,L)\{\bar{p}_{2,2}(L)[\bar{H}\bar{\Phi}S_{1312}(L-1)\Phi_V^T+\Phi_V^2S_{1322}(L-1)\Phi_V^T+\Phi_VS_{1332}(L-1)\Phi_V^T]$$
$$+\bar{p}(L)[\bar{H}\bar{\Phi}S_{2312}(L-1)\Phi_V^T+\Phi_V^2S_{2322}(L-1)\Phi_V^T+\Phi_VS_{2332}(L-1)\Phi_V^T]$$
$$-\bar{p}_{2,2}(L)[\bar{H}\bar{\Phi}S_{3312}(L-1)\Phi_V^T+\Phi_V^2S_{3322}(L-1)\Phi_V^T+\Phi_VS_{3332}(L-1)\Phi_V^T]\} \tag{3.11}$$

$$\bar{C}_{33}(k,L)=\bar{C}_{33}(k,L-1)\Phi_V^T+h(k,L,L)[\ 0 \quad p(L)I_m\ ]R_U(L)$$
$$-h(k,L,L)\{\bar{p}_{2,2}(L)[\bar{H}\bar{\Phi}S_{1313}(L-1)\Phi_V^T+\Phi_V^2S_{1323}(L-1)\Phi_V^T+\Phi_VS_{1333}(L-1)\Phi_V^T]$$
$$+\bar{p}(L)[\bar{H}\bar{\Phi}S_{2313}(L-1)\Phi_V^T+\Phi_V^2S_{2323}(L-1)\Phi_V^T+\Phi_VS_{2333}(L-1)\Phi_V^T]$$
$$-\bar{p}_{2,2}(L)[\bar{H}\bar{\Phi}S_{3313}(L-1)\Phi_V^T+\Phi_V^2S_{3323}(L-1)\Phi_V^T+\Phi_VS_{3333}(L-1)\Phi_V^T]\} \tag{3.12}$$

$$\bar{C}_{11}(L,L)=h(L,L,L)[\ p(L)I_m \quad 0\ ]\bar{H}K_{\bar{x}}(L,L)+\bar{\Phi}\tilde{B}_{11}(L-1)\bar{\Phi}^T$$
$$-h(L,L,L)\{\bar{p}_{2,2}(L)[\bar{H}\bar{\Phi}S_{1111}(L-1)\bar{\Phi}^T+\Phi_V^2S_{1121}(L-1)\bar{\Phi}^T+\Phi_VS_{1131}(L-1)\bar{\Phi}^T]$$
$$+\bar{p}(L)[\bar{H}\bar{\Phi}S_{2111}(L-1)\bar{\Phi}^T+\Phi_V^2S_{2121}(L-1)\bar{\Phi}^T+\Phi_VS_{2131}(L-1)\bar{\Phi}^T]$$
$$-\bar{p}_{2,2}(L)[\bar{H}\bar{\Phi}S_{3111}(L-1)\bar{\Phi}^T+\Phi_V^2S_{3121}(L-1)\bar{\Phi}^T+\Phi_VS_{3131}(L-1)\bar{\Phi}^T\}\ ,\quad \bar{C}_{11}(0,0)=0 \tag{3.13}$$

$$\bar{C}_{12}(L,L)=h(L,L,L)[\ p(L)I_m \quad 0\ ]\Phi_VK_V(L,L)+\bar{\Phi}\tilde{B}_{12}(L-1)\Phi_V^T$$
$$-h(L,L,L)\{\bar{p}_{2,2}(L)[\bar{H}\bar{\Phi}S_{1112}(L-1)\Phi_V^T+\Phi_V^2S_{1122}(L-1)\Phi_V^T+\Phi_VS_{1132}(L-1)\Phi_V^T]$$
$$+\bar{p}(L)[\bar{H}\bar{\Phi}S_{2112}(L-1)\Phi_V^T+\Phi_V^2S_{2122}(L-1)\Phi_V^T+\Phi_VS_{2132}(L-1)\Phi_V^T]$$
$$-\bar{p}_{2,2}(L)[\bar{H}\bar{\Phi}S_{3112}(L-1)\Phi_V^T+\Phi_V^2S_{3122}(L-1)\Phi_V^T+\Phi_VS_{3132}(L-1)\Phi_V^T\}\ ,\quad \bar{C}_{12}(0,0)=0 \tag{3.14}$$

$$\bar{C}_{13}(L,L)=h(L,L,L)[\ p(L)I_m \quad 0\ ]R_U(L)+\bar{\Phi}\tilde{B}_{13}(L-1)\Phi_V^T$$
$$-h(L,L,L)\{\bar{p}_{2,2}(L)[\bar{H}\bar{\Phi}S_{1113}(L-1)\Phi_V^T+\Phi_V^2S_{1123}(L-1)\Phi_V^T+\Phi_VS_{1133}(L-1)\Phi_V^T]$$
$$+\bar{p}(L)[\bar{H}\bar{\Phi}S_{2113}(L-1)\Phi_V^T+\Phi_V^2S_{2123}(L-1)\Phi_V^T+\Phi_VS_{2133}(L-1)\Phi_V^T]$$
$$-\bar{p}_{2,2}(L)[\bar{H}\bar{\Phi}S_{3113}(L-1)\Phi_V^T+\Phi_V^2S_{3123}(L-1)\Phi_V^T+\Phi_VS_{3133}(L-1)\Phi_V^T\}\ ,\quad \bar{C}_{13}(0,0)=0 \tag{3.15}$$

$$\begin{aligned}
&\bar{C}_{21}(L,L) = h(L,L,L)[\ 0 \quad I_m\ ]\bar{H}K_{\bar{x}}(L,L) + \bar{\Phi}\tilde{B}_{21}(L-1)\bar{\Phi}^T \\
&-h(L,L,L)\{\bar{p}_{2,2}(L)[\bar{H}\bar{\Phi}S_{1211}(L-1)\bar{\Phi}^T + \Phi_V^2 S_{1221}(L-1)\bar{\Phi}^T + \Phi_V S_{1231}(L-1)\bar{\Phi}^T] \\
&+\bar{p}(L)[\bar{H}\bar{\Phi}S_{2211}(L-1)\bar{\Phi}^T + \Phi_V^2 S_{2221}(L-1)\bar{\Phi}^T + \Phi_V S_{2231}(L-1)\bar{\Phi}^T] \\
&-\bar{p}_{2,2}(L)[\bar{H}\bar{\Phi}S_{3211}(L-1)\bar{\Phi}^T + \Phi_V^2 S_{3221}(L-1)\bar{\Phi}^T + \Phi_V S_{3231}(L-1)\bar{\Phi}^T\}\ , \quad \bar{C}_{21}(0,0) = 0
\end{aligned} \tag{3.16}$$

$$\begin{aligned}
&\bar{C}_{22}(L,L) = h(L,L,L)[\ 0 \quad I_m\ ]\Phi_V K_V(L,L) + \bar{\Phi}\tilde{B}_{22}(L-1)\Phi_V^T \\
&-h(L,L,L)\{\bar{p}_{2,2}(L)[\bar{H}\bar{\Phi}S_{1212}(L-1)\Phi_V^T + \Phi_V^2 S_{1222}(L-1)\Phi_V^T + \Phi_V S_{1232}(L-1)\Phi_V^T] \\
&+\bar{p}(L)[\bar{H}\bar{\Phi}S_{2212}(L-1)\Phi_V^T + \Phi_V^2 S_{2222}(L-1)\Phi_V^T + \Phi_V S_{2232}(L-1)\Phi_V^T] \\
&-\bar{p}_{2,2}(L)[\bar{H}\bar{\Phi}S_{3212}(L-1)\Phi_V^T + \Phi_V^2 S_{3222}(L-1)\Phi_V^T + \Phi_V S_{3232}(L-1)\Phi_V^T\}\ , \quad \bar{C}_{22}(0,0) = 0
\end{aligned} \tag{3.17}$$

$$\begin{aligned}
&\bar{C}_{23}(L,L) = h(L,L,L)[\ 0 \quad I_m\ ]R_U(L) + \bar{\Phi}\tilde{B}_{23}(L-1)\Phi_V^T \\
&-h(L,L,L)\{\bar{p}_{2,2}(L)[\bar{H}\bar{\Phi}S_{1213}(L-1)\Phi_V^T + \Phi_V^2 S_{1223}(L-1)\Phi_V^T + \Phi_V S_{1233}(L-1)\Phi_V^T] \\
&+\bar{p}(L)[\bar{H}\bar{\Phi}S_{2213}(L-1)\Phi_V^T + \Phi_V^2 S_{2223}(L-1)\Phi_V^T + \Phi_V S_{2233}(L-1)\Phi_V^T] \\
&-\bar{p}_{2,2}(L)[\bar{H}\bar{\Phi}S_{3213}(L-1)\Phi_V^T + \Phi_V^2 S_{3223}(L-1)\Phi_V^T + \Phi_V S_{3233}(L-1)\Phi_V^T\}\ , \quad \bar{C}_{23}(0,0) = 0
\end{aligned} \tag{3.18}$$

$$\begin{aligned}
&\bar{C}_{31}(L,L) = h(L,L,L)[\ 0 \quad p(L)I_m\ ]\bar{H}K_{\bar{x}}(L,L) + \bar{\Phi}\tilde{B}_{31}(L-1)\bar{\Phi}^T \\
&-h(L,L,L)\{\bar{p}_{2,2}(L)[\bar{H}\bar{\Phi}S_{1311}(L-1)\bar{\Phi}^T + \Phi_V^2 S_{1321}(L-1)\bar{\Phi}^T + \Phi_V S_{1331}(L-1)\bar{\Phi}^T] \\
&+\bar{p}(L)[\bar{H}\bar{\Phi}S_{2311}(L-1)\bar{\Phi}^T + \Phi_V^2 S_{2321}(L-1)\bar{\Phi}^T + \Phi_V S_{2331}(L-1)\bar{\Phi}^T] \\
&-\bar{p}_{2,2}(L)[\bar{H}\bar{\Phi}S_{3311}(L-1)\bar{\Phi}^T + \Phi_V^2 S_{3321}(L-1)\bar{\Phi}^T + \Phi_V S_{3331}(L-1)\bar{\Phi}^T\}\ , \quad \bar{C}_{31}(0,0) = 0
\end{aligned} \tag{3.19}$$

$$\begin{aligned}
&\bar{C}_{32}(L,L) = h(L,L,L)[\ 0 \quad p(L)I_m\ ]\Phi_V K_V(L,L) + \bar{\Phi}\tilde{B}_{32}(L-1)\Phi_V^T \\
&-h(L,L,L)\{\bar{p}_{2,2}(L)[\bar{H}\bar{\Phi}S_{1312}(L-1)\Phi_V^T + \Phi_V^2 S_{1322}(L-1)\Phi_V^T + \Phi_V S_{1332}(L-1)\Phi_V^T] \\
&+\bar{p}(L)[\bar{H}\bar{\Phi}S_{2312}(L-1)\Phi_V^T + \Phi_V^2 S_{2322}(L-1)\Phi_V^T + \Phi_V S_{2332}(L-1)\Phi_V^T] \\
&-\bar{p}_{2,2}(L)[\bar{H}\bar{\Phi}S_{3312}(L-1)\Phi_V^T + \Phi_V^2 S_{3322}(L-1)\Phi_V^T + \Phi_V S_{3332}(L-1)\Phi_V^T\}\ , \quad \bar{C}_{32}(0,0) = 0
\end{aligned} \tag{3.20}$$

$$\begin{aligned}
&\bar{C}_{33}(L,L) = h(L,L,L)[\ 0 \quad p(L)I_m\ ]R_U(L) + \bar{\Phi}\tilde{B}_{33}(L-1)\Phi_V^T \\
&-h(L,L,L)\{\bar{p}_{2,2}(L)[\bar{H}\bar{\Phi}S_{1313}(L-1)\Phi_V^T + \Phi_V^2 S_{1323}(L-1)\Phi_V^T + \Phi_V S_{1333}(L-1)\Phi_V^T] \\
&+\bar{p}(L)[\bar{H}\bar{\Phi}S_{2313}(L-1)\Phi_V^T + \Phi_V^2 S_{2323}(L-1)\Phi_V^T + \Phi_V S_{2333}(L-1)\Phi_V^T] \\
&-\bar{p}_{2,2}(L)[\bar{H}\bar{\Phi}S_{3313}(L-1)\Phi_V^T + \Phi_V^2 S_{3323}(L-1)\Phi_V^T + \Phi_V S_{3333}(L-1)\Phi_V^T\}\ , \quad \bar{C}_{33}(0,0) = 0
\end{aligned} \tag{3.21}$$

$$
\begin{aligned}
\Gamma(L) &= \Lambda(L) - \bar{p}_{2,2}(L)\{[\bar{H}\bar{\Phi}S_{1111}(L-1)\bar{\Phi}^T\bar{H}^T + \bar{H}\bar{\Phi}S_{1112}(L-1)(\Phi_V^T)^2 \\
&+\bar{H}\bar{\Phi}S_{1113}(L-1)(\Phi_V^T) + \Phi_V^2 S_{1121}(L-1)\bar{\Phi}^T\bar{H}^T + \Phi_V^2 S_{1122}(L-1)(\Phi_V^T)^2 \\
&+\Phi_V^2 S_{1123}(L-1)(\Phi_V^T) + \Phi_V S_{1131}(L-1)\bar{\Phi}^T\bar{H}^T + \Phi_V S_{1132}(L-1)(\Phi_V^T)^2 \\
&+\Phi_V S_{1133}(L-1)(\Phi_V^T)]\bar{p}_{2,2}^T(L) + [\bar{H}\bar{\Phi}S_{1211}(L-1)\bar{\Phi}^T\bar{H}^T + \bar{H}\bar{\Phi}S_{1212}(L-1)(\Phi_V^T)^2 \\
&+\bar{H}\bar{\Phi}S_{1213}(L-1)(\Phi_V^T) + \Phi_V^2 S_{1221}(L-1)\bar{\Phi}^T\bar{H}^T + \Phi_V^2 S_{1222}(L-1)(\Phi_V^T)^2 \\
&+\Phi_V^2 S_{1223}(L-1)(\Phi_V^T) + \Phi_V S_{1231}(L-1)\bar{\Phi}^T\bar{H}^T + \Phi_V S_{1232}(L-1)(\Phi_V^T)^2 \\
&+\Phi_V S_{1233}(L-1)(\Phi_V^T)]\bar{p}^T(L) - [\bar{H}\bar{\Phi}S_{1311}(L-1)\bar{\Phi}^T\bar{H}^T + \bar{H}\bar{\Phi}S_{1312}(L-1)(\Phi_V^T)^2 \\
&+\bar{H}\bar{\Phi}S_{1313}(L-1)(\Phi_V^T) + \Phi_V^2 S_{1321}(L-1)\bar{\Phi}^T\bar{H}^T + \Phi_V^2 S_{1322}(L-1)(\Phi_V^T)^2 \\
&+\Phi_V^2 S_{1323}(L-1)(\Phi_V^T) + \Phi_V S_{1331}(L-1)\bar{\Phi}^T\bar{H}^T + \Phi_V S_{1332}(L-1)(\Phi_V^T)^2 \\
&+\Phi_V S_{1333}(L-1)(\Phi_V^T)]\bar{p}_{2,2}^T(L) - \bar{p}(L)\{[\bar{H}\bar{\Phi}S_{2111}(L-1)\bar{\Phi}^T\bar{H}^T \\
&+\bar{H}\bar{\Phi}S_{2112}(L-1)(\Phi_V^T)^2 + \bar{H}\bar{\Phi}S_{2113}(L-1)(\Phi_V^T) + \Phi_V^2 S_{2121}(L-1)\bar{\Phi}^T\bar{H}^T \\
&+\Phi_V^2 S_{2122}(L-1)(\Phi_V^T)^2 + \Phi_V^2 S_{2123}(L-1)(\Phi_V^T) + \Phi_V S_{2131}(L-1)\bar{\Phi}^T\bar{H}^T \\
&+\Phi_V S_{2132}(L-1)(\Phi_V^T)^2 + \Phi_V S_{2133}(L-1)(\Phi_V^T)]\bar{p}_{2,2}^T(L) \\
&+[\bar{H}\bar{\Phi}S_{2211}(L-1)\bar{\Phi}^T\bar{H}^T + \bar{H}\bar{\Phi}S_{2212}(L-1)(\Phi_V^T)^2 + \bar{H}\bar{\Phi}S_{2213}(L-1)(\Phi_V^T) \\
&+\Phi_V^2 S_{2221}(L-1)\bar{\Phi}^T\bar{H}^T + \Phi_V^2 S_{2222}(L-1)(\Phi_V^T)^2 + \Phi_V^2 S_{2223}(L-1)(\Phi_V^T) \\
&+\Phi_V S_{2231}(L-1)\bar{\Phi}^T\bar{H}^T + \Phi_V S_{2232}(L-1)(\Phi_V^T)^2 + \Phi_V S_{2233}(L-1)(\Phi_V^T)]\bar{p}^T(L) \\
&-[\bar{H}\bar{\Phi}S_{2311}(L-1)\bar{\Phi}^T\bar{H}^T + \bar{H}\bar{\Phi}S_{2312}(L-1)(\Phi_V^T)^2 + \bar{H}\bar{\Phi}S_{2313}(L-1)(\Phi_V^T) \\
&+\Phi_V^2 S_{2321}(L-1)\bar{\Phi}^T\bar{H}^T + \Phi_V^2 S_{2322}(L-1)(\Phi_V^T)^2 \\
&+\Phi_V^2 S_{2323}(L-1)(\Phi_V^T) + \Phi_V S_{2331}(L-1)\bar{\Phi}^T\bar{H}^T + \Phi_V S_{2332}(L-1)(\Phi_V^T)^2 \\
&+\Phi_V S_{2333}(L-1)(\Phi_V^T)]\bar{p}_{2,2}^T(L) + \bar{p}_{2,2}(L)\{[\bar{H}\bar{\Phi}S_{3111}(L-1)\bar{\Phi}^T\bar{H}^T \\
&+\bar{H}\bar{\Phi}S_{3112}(L-1)(\Phi_V^T)^2 + \bar{H}\bar{\Phi}S_{3113}(L-1)(\Phi_V^T) + \Phi_V^2 S_{3121}(L-1)\bar{\Phi}^T\bar{H}^T \\
&+\Phi_V^2 S_{3122}(L-1)(\Phi_V^T)^2 + \Phi_V^2 S_{3123}(L-1)(\Phi_V^T) + \Phi_V S_{3131}(L-1)\bar{\Phi}^T\bar{H}^T \\
&+\Phi_V S_{3132}(L-1)(\Phi_V^T)^2 + \Phi_V S_{3133}(L-1)(\Phi_V^T)]\bar{p}_{2,2}^T(L) \\
&+[\bar{H}\bar{\Phi}S_{3211}(L-1)\bar{\Phi}^T\bar{H}^T + \bar{H}\bar{\Phi}S_{3212}(L-1)(\Phi_V^T)^2 \\
&+\bar{H}\bar{\Phi}S_{3213}(L-1)(\Phi_V^T) + \Phi_V^2 S_{3221}(L-1)\bar{\Phi}^T\bar{H}^T + \Phi_V^2 S_{3222}(L-1)(\Phi_V^T)^2 \\
&+\Phi_V^2 S_{3223}(L-1)(\Phi_V^T) + \Phi_V S_{3231}(L-1)\bar{\Phi}^T\bar{H}^T + \Phi_V S_{3232}(L-1)(\Phi_V^T)^2 \\
&+\Phi_V S_{3233}(L-1)(\Phi_V^T)]\bar{p}^T(L) - [\bar{H}\bar{\Phi}S_{3311}(L-1)\bar{\Phi}^T\bar{H}^T + \bar{H}\bar{\Phi}S_{3312}(L-1)(\Phi_V^T)^2 \\
&+\bar{H}\bar{\Phi}S_{3313}(L-1)(\Phi_V^T) + \Phi_V^2 S_{3321}(L-1)\bar{\Phi}^T\bar{H}^T + \Phi_V^2 S_{3322}(L-1)(\Phi_V^T)^2 \\
&+\Phi_V^2 S_{3323}(L-1)(\Phi_V^T) + \Phi_V S_{3331}(L-1)\bar{\Phi}^T\bar{H}^T + \Phi_V S_{3332}(L-1)(\Phi_V^T)^2 \\
&+\Phi_V S_{3333}(L-1)(\Phi_V^T)]\bar{p}_{2,2}^T(L)\},
\end{aligned}
\tag{3.22}
$$

$$
\begin{aligned}
\Lambda(L) &= \bar{R}_U(L) + \begin{bmatrix} p(L)I_m & 0 \end{bmatrix}(\bar{H}K_V(L,L)\bar{H}^T + \Phi_V K_V(L,L)\Phi_V^T)\begin{bmatrix} I_m & 0 \end{bmatrix}^T \\
&+\begin{bmatrix} 0 & (1-p(L))I_m \end{bmatrix}(\bar{H}K_V(L,L)\bar{H}^T + \Phi_V K_V(L,L)\Phi_V^T)\begin{bmatrix} 0 & I_m \end{bmatrix}^T
\end{aligned}
\tag{3.23}
$$

$$\bar{R}_U(L) = E_\gamma[\bar{\gamma}(s)R_U(L)\bar{\gamma}^T(L)] \tag{3.24}$$

$$\bar{O}_{11}(L) = \bar{\Phi}\bar{O}_{11}(L-1) + \tilde{J}_{11}(L,L)\{y(L) - \bar{p}_{2,2}(L)[\bar{H}\bar{\Phi}\bar{O}_{11}(L-1) + \Phi_V^2\bar{O}_{12}(L-1) + \Phi_V\bar{O}_{13}(L-1)] - \bar{p}(L)[\bar{H}\bar{\Phi}\bar{O}_{21}(L-1) + \Phi_V^2\bar{O}_{22}(L-1) + \Phi_V\bar{O}_{23}(L-1)] + \bar{p}_{2,2}(L)[\bar{H}\bar{\Phi}\bar{O}_{31}(L-1) + \Phi_V^2\bar{O}_{32}(L-1) + \Phi_V\bar{O}_{33}(L-1)]\}, \quad \bar{O}_{11}(0) = 0 \tag{3.25}$$

$$\bar{O}_{12}(L) = \Phi_V\bar{O}_{12}(L-1) + \tilde{J}_{12}(L,L)\{y(L) - \bar{p}_{2,2}(L)[\bar{H}\bar{\Phi}\bar{O}_{11}(L-1) + \Phi_V^2\bar{O}_{12}(L-1) + \Phi_V\bar{O}_{13}(L-1)] - \bar{p}(L)[\bar{H}\bar{\Phi}\bar{O}_{21}(L-1) + \Phi_V^2\bar{O}_{22}(L-1) + \Phi_V\bar{O}_{23}(L-1)] + \bar{p}_{2,2}(L)[\bar{H}\bar{\Phi}\bar{O}_{31}(L-1) + \Phi_V^2\bar{O}_{32}(L-1) + \Phi_V\bar{O}_{33}(L-1)]\}, \quad \bar{O}_{12}(0) = 0 \tag{3.26}$$

$$\bar{O}_{13}(L) = \Phi_V\bar{O}_{13}(L-1) + \tilde{J}_{13}(L,L)\{y(L) - \bar{p}_{2,2}(L)[\bar{H}\bar{\Phi}\bar{O}_{11}(L-1) + \Phi_V^2\bar{O}_{12}(L-1) + \Phi_V\bar{O}_{13}(L-1)] - \bar{p}(L)[\bar{H}\bar{\Phi}\bar{O}_{21}(L-1) + \Phi_V^2\bar{O}_{22}(L-1) + \Phi_V\bar{O}_{23}(L-1)] + \bar{p}_{2,2}(L)[\bar{H}\bar{\Phi}\bar{O}_{31}(L-1) + \Phi_V^2\bar{O}_{32}(L-1) + \Phi_V\bar{O}_{33}(L-1)]\}, \quad \bar{O}_{13}(0) = 0 \tag{3.27}$$

$$\bar{O}_{21}(L) = \bar{\Phi}\bar{O}_{21}(L-1) + \tilde{J}_{21}(L,L)\{y(L) - \bar{p}_{2,2}(L)[\bar{H}\bar{\Phi}\bar{O}_{11}(L-1) + \Phi_V^2\bar{O}_{12}(L-1) + \Phi_V\bar{O}_{13}(L-1)] - \bar{p}(L)[\bar{H}\bar{\Phi}\bar{O}_{21}(L-1) + \Phi_V^2\bar{O}_{22}(L-1) + \Phi_V\bar{O}_{23}(L-1)] + \bar{p}_{2,2}(L)[\bar{H}\bar{\Phi}\bar{O}_{31}(L-1) + \Phi_V^2\bar{O}_{32}(L-1) + \Phi_V\bar{O}_{33}(L-1)]\}, \quad \bar{O}_{21}(0) = 0 \tag{3.28}$$

$$\bar{O}_{22}(L) = \Phi_V\bar{O}_{22}(L-1) + \tilde{J}_{22}(L,L)\{y(L) - \bar{p}_{2,2}(L)[\bar{H}\bar{\Phi}\bar{O}_{11}(L-1) + \Phi_V^2\bar{O}_{12}(L-1) + \Phi_V\bar{O}_{13}(L-1)] - \bar{p}(L)[\bar{H}\bar{\Phi}\bar{O}_{21}(L-1) + \Phi_V^2\bar{O}_{22}(L-1) + \Phi_V\bar{O}_{23}(L-1)] + \bar{p}_{2,2}(L)[\bar{H}\bar{\Phi}\bar{O}_{31}(L-1) + \Phi_V^2\bar{O}_{32}(L-1) + \Phi_V\bar{O}_{33}(L-1)]\}, \quad \bar{O}_{22}(0) = 0 \tag{3.29}$$

$$\bar{O}_{23}(L) = \Phi_V\bar{O}_{23}(L-1) + \tilde{J}_{23}(L,L)\{y(L) - \bar{p}_{2,2}(L)[\bar{H}\bar{\Phi}\bar{O}_{11}(L-1) + \Phi_V^2\bar{O}_{12}(L-1) + \Phi_V\bar{O}_{13}(L-1)] - \bar{p}(L)[\bar{H}\bar{\Phi}\bar{O}_{21}(L-1) + \Phi_V^2\bar{O}_{22}(L-1) + \Phi_V\bar{O}_{23}(L-1)] + \bar{p}_{2,2}(L)[\bar{H}\bar{\Phi}\bar{O}_{31}(L-1) + \Phi_V^2\bar{O}_{32}(L-1) + \Phi_V\bar{O}_{33}(L-1)]\}, \quad \bar{O}_{13}(0) = 0 \tag{3.30}$$

$$\bar{O}_{31}(L) = \bar{\Phi}\bar{O}_{31}(L-1) + \tilde{J}_{31}(L,L)\{y(L) - \bar{p}_{2,2}(L)[\bar{H}\bar{\Phi}\bar{O}_{11}(L-1) + \Phi_V^2\bar{O}_{12}(L-1) + \Phi_V\bar{O}_{13}(L-1)] - \bar{p}(L)[\bar{H}\bar{\Phi}\bar{O}_{21}(L-1) + \Phi_V^2\bar{O}_{22}(L-1) + \Phi_V\bar{O}_{23}(L-1)] + \bar{p}_{2,2}(L)[\bar{H}\bar{\Phi}\bar{O}_{31}(L-1) + \Phi_V^2\bar{O}_{32}(L-1) + \Phi_V\bar{O}_{33}(L-1)]\}, \quad \bar{O}_{31}(0) = 0 \tag{3.31}$$

$$\bar{O}_{32}(L) = \Phi_V\bar{O}_{32}(L-1) + \tilde{J}_{32}(L,L)\{y(L) - \bar{p}_{2,2}(L)[\bar{H}\bar{\Phi}\bar{O}_{11}(L-1) + \Phi_V^2\bar{O}_{12}(L-1) + \Phi_V\bar{O}_{13}(L-1)] - \bar{p}(L)[\bar{H}\bar{\Phi}\bar{O}_{21}(L-1) + \Phi_V^2\bar{O}_{22}(L-1) + \Phi_V\bar{O}_{23}(L-1)] + \bar{p}_{2,2}(L)[\bar{H}\bar{\Phi}\bar{O}_{31}(L-1) + \Phi_V^2\bar{O}_{32}(L-1) + \Phi_V\bar{O}_{33}(L-1)]\}, \quad \bar{O}_{32}(0) = 0 \tag{3.32}$$

$$\begin{aligned}\bar{O}_{33}(L) &= \Phi_V \bar{O}_{33}(L-1) + \tilde{J}_{33}(L,L)\{y(L) - \bar{p}_{2,2}(L)[\bar{H}\bar{\Phi}\bar{O}_{11}(L-1) + \Phi_V^2 \bar{O}_{12}(L-1) \\ &+ \Phi_V \bar{O}_{13}(L-1)] - \bar{p}(L)[\bar{H}\bar{\Phi}\bar{O}_{21}(L-1) + \Phi_V^2 \bar{O}_{22}(L-1) + \Phi_V \bar{O}_{23}(L-1)] \\ &+ \bar{p}_{2,2}(L)[\bar{H}\bar{\Phi}\bar{O}_{31}(L-1) + \Phi_V^2 \bar{O}_{32}(L-1) + \Phi_V \bar{O}_{33}(L-1)]\}, \quad \bar{O}_{33}(0) = 0\end{aligned} \tag{3.33}$$

$$\begin{aligned}\tilde{J}_{11}(L,L) &= \{K_{\bar{x}}(L,L)\bar{H}^T \begin{bmatrix} p(L)I_m \\ 0 \end{bmatrix} - [\bar{\Phi}S_{1111}(L-1)\bar{\Phi}^T\bar{H}^T \\ &+ \bar{\Phi}S_{1112}(L-1)(\Phi_V^T)^2 + \bar{\Phi}S_{1113}(L-1)\Phi_V^T]\bar{p}_{2,2}^T(L) - [\bar{\Phi}S_{1211}(L-1)\bar{\Phi}^T\bar{H}^T \\ &+ \bar{\Phi}S_{1212}(L-1)(\Phi_V^T)^2 + \bar{\Phi}S_{1213}(L-1)\Phi_V^T]\bar{p}^T(L) + [\bar{\Phi}S_{1311}(L-1)\bar{\Phi}^T\bar{H}^T \\ &+ \bar{\Phi}S_{1312}(L-1)(\Phi_V^T)^2 + \bar{\Phi}S_{1313}(L-1)\Phi_V^T]\bar{p}_{2,2}^T(L)\}\Gamma^{-1}(L)\end{aligned} \tag{3.34}$$

$$\begin{aligned}\tilde{J}_{12}(L,L) &= \{K_V(L,L)\Phi_V^T \begin{bmatrix} p(L)I_m \\ 0 \end{bmatrix} - [\Phi_V S_{1121}(L-1)\bar{\Phi}^T\bar{H}^T \\ &+ \Phi_V S_{1122}(L-1)(\Phi_V^T)^2 + \Phi_V S_{1123}(L-1)\Phi_V^T]\bar{p}_{2,2}^T(L) - [\Phi_V S_{1221}(L-1)\bar{\Phi}^T\bar{H}^T \\ &+ \Phi_V S_{1222}(L-1)(\Phi_V^T)^2 + \Phi_V S_{1223}(L-1)\Phi_V^T]\bar{p}^T(L) + [\Phi_V S_{1321}(L-1)\bar{\Phi}^T\bar{H}^T \\ &+ \Phi_V S_{1322}(L-1)(\Phi_V^T)^2 + \Phi_V S_{1323}(L-1)\Phi_V^T]\bar{p}_{2,2}^T(L)\}\Gamma^{-1}(L)\end{aligned} \tag{3.35}$$

$$\begin{aligned}\tilde{J}_{13}(L,L) &= \{\bar{R}_U(L) \begin{bmatrix} p(L)I_m \\ 0 \end{bmatrix} - [\Phi_V S_{1131}(L-1)\bar{\Phi}^T\bar{H}^T \\ &+ \Phi_V S_{1132}(L-1)(\Phi_V^T)^2 + \Phi_V S_{1133}(L-1)\Phi_V^T]\bar{p}_{2,2}^T(L) - [\Phi_V S_{1231}(L-1)\bar{\Phi}^T\bar{H}^T \\ &+ \Phi_V S_{1232}(L-1)(\Phi_V^T)^2 + \Phi_V S_{1233}(L-1)\Phi_V^T]\bar{p}^T(L) + [\Phi_V S_{1331}(L-1)\bar{\Phi}^T\bar{H}^T \\ &+ \Phi_V S_{1332}(L-1)(\Phi_V^T)^2 + \Phi_V S_{1333}(L-1)\Phi_V^T]\bar{p}_{2,2}^T(L)\}\Gamma^{-1}(L)\end{aligned} \tag{3.36}$$

$$\begin{aligned}\tilde{J}_{21}(L,L) &= \{K_{\bar{x}}(L,L)\bar{H}^T \begin{bmatrix} 0 \\ I_m \end{bmatrix} - [\bar{\Phi}S_{2111}(L-1)\bar{\Phi}^T\bar{H}^T \\ &+ \bar{\Phi}S_{2112}(L-1)(\Phi_V^T)^2 + \bar{\Phi}S_{2113}(L-1)\Phi_V^T]\bar{p}_{2,2}^T(L) - [\bar{\Phi}S_{2211}(L-1)\bar{\Phi}^T\bar{H}^T \\ &+ \bar{\Phi}S_{2212}(L-1)(\Phi_V^T)^2 + \bar{\Phi}S_{2213}(L-1)\Phi_V^T]\bar{p}^T(L) + [\bar{\Phi}S_{2311}(L-1)\bar{\Phi}^T\bar{H}^T \\ &+ \bar{\Phi}S_{2312}(L-1)(\Phi_V^T)^2 + \bar{\Phi}S_{2313}(L-1)\Phi_V^T]\bar{p}_{2,2}^T(L)\}\Gamma^{-1}(L)\end{aligned} \tag{3.37}$$

$$\begin{aligned}\tilde{J}_{22}(L,L) &= \{K_V(L,L)\Phi_V^T \begin{bmatrix} 0 \\ I_m \end{bmatrix} - [\Phi_V S_{2121}(L-1)\bar{\Phi}^T\bar{H}^T \\ &+ \Phi_V S_{2122}(L-1)(\Phi_V^T)^2 + \Phi_V S_{2123}(L-1)\Phi_V^T]\bar{p}_{2,2}^T(L) - [\Phi_V S_{2221}(L-1)\bar{\Phi}^T\bar{H}^T \\ &+ \Phi_V S_{2222}(L-1)(\Phi_V^T)^2 + \Phi_V S_{2223}(L-1)\Phi_V^T]\bar{p}^T(L) + [\Phi_V S_{2321}(L-1)\bar{\Phi}^T\bar{H}^T \\ &+ \Phi_V S_{2322}(L-1)(\Phi_V^T)^2 + \Phi_V S_{2323}(L-1)\Phi_V^T]\bar{p}_{2,2}^T(L)\}\Gamma^{-1}(L)\end{aligned} \tag{3.38}$$

$$
\begin{aligned}
\tilde{J}_{23}(L,L) &= \{\bar{R}_U(L)\begin{bmatrix} 0 \\ I_m \end{bmatrix} - [\Phi_V S_{2131}(L-1)\bar{\Phi}^T\bar{H}^T \\
&+\Phi_V S_{2132}(L-1)(\Phi_V^T)^2 + \Phi_V S_{2133}(L-1)\Phi_V^T]\bar{p}_{2,2}^T(L) - [\Phi_V S_{2231}(L-1)\bar{\Phi}^T\bar{H}^T \\
&+\Phi_V S_{2232}(L-1)(\Phi_V^T)^2 + \Phi_V S_{2233}(L-1)\Phi_V^T]\bar{p}^T(L) + [\Phi_V S_{2331}(L-1)\bar{\Phi}^T\bar{H}^T \\
&+\Phi_V S_{2332}(L-1)(\Phi_V^T)^2 + \Phi_V S_{2333}(L-1)\Phi_V^T]\bar{p}_{2,2}^T(L)\}\Gamma^{-1}(L)
\end{aligned} \tag{3.39}
$$

$$
\begin{aligned}
\tilde{J}_{31}(L,L) &= \{K_{\bar{x}}(L,L)\bar{H}^T\begin{bmatrix} 0 \\ p(L)I_m \end{bmatrix} - [\bar{\Phi}S_{3111}(L-1)\bar{\Phi}^T\bar{H}^T \\
&+\bar{\Phi}S_{3112}(L-1)(\Phi_V^T)^2 + \bar{\Phi}S_{3113}(L-1)\Phi_V^T]\bar{p}_{2,2}^T(L) - [\bar{\Phi}S_{3211}(L-1)\bar{\Phi}^T\bar{H}^T \\
&+\bar{\Phi}S_{3212}(L-1)(\Phi_V^T)^2 + \bar{\Phi}S_{3213}(L-1)\Phi_V^T]\bar{p}^T(L) + [\bar{\Phi}S_{3311}(L-1)\bar{\Phi}^T\bar{H}^T \\
&+\bar{\Phi}S_{3312}(L-1)(\Phi_V^T)^2 + \bar{\Phi}S_{3313}(L-1)\Phi_V^T]\bar{p}_{2,2}^T(L)\}\Gamma^{-1}(L)
\end{aligned} \tag{3.40}
$$

$$
\begin{aligned}
\tilde{J}_{32}(L,L) &= \{K_V(L,L)\Phi_V^T\begin{bmatrix} 0 \\ p(L)I_m \end{bmatrix} - [\Phi_V S_{3121}(L-1)\bar{\Phi}^T\bar{H}^T \\
&+\Phi_V S_{3122}(L-1)(\Phi_V^T)^2 + \Phi_V S_{3123}(L-1)\Phi_V^T]\bar{p}_{2,2}^T(L) - [\Phi_V S_{3221}(L-1)\bar{\Phi}^T\bar{H}^T \\
&+\Phi_V S_{3222}(L-1)(\Phi_V^T)^2 + \Phi_V S_{3223}(L-1)\Phi_V^T]\bar{p}^T(L) + [\Phi_V S_{3321}(L-1)\bar{\Phi}^T\bar{H}^T \\
&+\Phi_V S_{3322}(L-1)(\Phi_V^T)^2 + \Phi_V S_{3323}(L-1)\Phi_V^T]\bar{p}_{2,2}^T(L)\}\Gamma^{-1}(L)
\end{aligned} \tag{3.41}
$$

$$
\begin{aligned}
\tilde{J}_{33}(L,L) &= \{\bar{R}_U(L)\begin{bmatrix} 0 \\ p(L)I_m \end{bmatrix} - [\Phi_V S_{3131}(L-1)\bar{\Phi}^T\bar{H}^T \\
&+\Phi_V S_{3132}(L-1)(\Phi_V^T)^2 + \Phi_V S_{3133}(L-1)\Phi_V^T]\bar{p}_{2,2}^T(L) - [\Phi_V S_{3231}(L-1)\bar{\Phi}^T\bar{H}^T \\
&+\Phi_V S_{3232}(L-1)(\Phi_V^T)^2 + \Phi_V S_{3233}(L-1)\Phi_V^T]\bar{p}^T(L) + [\Phi_V S_{3331}(L-1)\bar{\Phi}^T\bar{H}^T \\
&+\Phi_V S_{3332}(L-1)(\Phi_V^T)^2 + \Phi_V S_{3333}(L-1)\Phi_V^T]\bar{p}_{2,2}^T(L)\}\Gamma^{-1}(L)
\end{aligned} \tag{3.42}
$$

$$
\begin{aligned}
S_{1111}(L) &= \tilde{J}_{11}(L,L)[\ p(L)I_m \quad 0\ ]\bar{H}K_{\bar{x}}(L,L) + \bar{\Phi}S_{1111}(L-1)\bar{\Phi}^T \\
&-\tilde{J}_{11}(L,L)\{\bar{p}_{2,2}(L)[\bar{H}\bar{\Phi}S_{1111}(L-1)\bar{\Phi}^T + \Phi_V^2 S_{1121}(L-1)\bar{\Phi}^T + \Phi_V S_{1131}(L-1)\bar{\Phi}^T] \\
&+\bar{p}(L)[\bar{H}\bar{\Phi}S_{2111}(L-1)\bar{\Phi}^T + \Phi_V^2 S_{2121}(L-1)\bar{\Phi}^T + \Phi_V S_{2131}(L-1)\bar{\Phi}^T] \\
&-\bar{p}_{2,2}(L)[\bar{H}\bar{\Phi}S_{3111}(L-1)\bar{\Phi}^T + \Phi_V^2 S_{3121}(L-1)\bar{\Phi}^T + \Phi_V S_{3131}(L-1)\bar{\Phi}^T]\}\ , \quad S_{1111}(0) = 0
\end{aligned} \tag{3.43}
$$

$$
\begin{aligned}
S_{1112}(L) &= \tilde{J}_{11}(L,L)[\ p(L)I_m \quad 0\ ]\Phi_V K_V(L,L) + \bar{\Phi}S_{1112}(L-1)\bar{\Phi}^T \\
&-\tilde{J}_{11}(L,L)\{\bar{p}_{2,2}(L)[\bar{H}\bar{\Phi}S_{1112}(L-1)\Phi_V^T + \Phi_V^2 S_{1122}(L-1)\Phi_V^T + \Phi_V S_{1132}(L-1)\Phi_V^T] \\
&+\bar{p}(L)[\bar{H}\bar{\Phi}S_{2112}(L-1)\Phi_V^T + \Phi_V^2 S_{2122}(L-1)\Phi_V^T + \Phi_V S_{2132}(L-1)\Phi_V^T] \\
&-\bar{p}_{2,2}(L)[\bar{H}\bar{\Phi}S_{3112}(L-1)\Phi_V^T + \Phi_V^2 S_{3122}(L-1)\Phi_V^T + \Phi_V S_{3132}(L-1)\Phi_V^T]\}\ , \quad S_{1112}(0) = 0
\end{aligned} \tag{3.44}
$$

$$
\begin{aligned}
&S_{1113}(L) = \tilde{J}_{11}(L,L)[\ p(L)I_m \quad 0\ ]R_U(L) + \bar{\Phi}S_{1113}(L-1)\Phi_V^T \\
&-\tilde{J}_{11}(L,L)\{\bar{p}_{2,2}(L)[\bar{H}\bar{\Phi}S_{1113}(L-1)\Phi_V^T + \Phi_V^2 S_{1123}(L-1)\Phi_V^T + \Phi_V S_{1133}(L-1)\Phi_V^T] \\
&+\bar{p}(L)[\bar{H}\bar{\Phi}S_{2113}(L-1)\Phi_V^T + \Phi_V^2 S_{2123}(L-1)\Phi_V^T + \Phi_V S_{2133}(L-1)\Phi_V^T] \\
&-\bar{p}_{2,2}(L)[\bar{H}\bar{\Phi}S_{3113}(L-1)\Phi_V^T + \Phi_V^2 S_{3123}(L-1)\Phi_V^T + \Phi_V S_{3133}(L-1)\Phi_V^T]\} \ , \quad S_{1113}(0) = 0
\end{aligned} \tag{3.45}
$$

$$
\begin{aligned}
&S_{1121}(L) = \tilde{J}_{12}(L,L)[\ p(L)I_m \quad 0\ ]\bar{H}K_{\bar{x}}(L,L) + \Phi_V S_{1121}(L-1)\bar{\Phi}^T \\
&-\tilde{J}_{12}(L,L)\{\bar{p}_{2,2}(L)[\bar{H}\bar{\Phi}S_{1111}(L-1)\bar{\Phi}^T + \Phi_V^2 S_{1121}(L-1)\bar{\Phi}^T + \Phi_V S_{1131}(L-1)\bar{\Phi}^T] \\
&+\bar{p}(L)[\bar{H}\bar{\Phi}S_{2111}(L-1)\bar{\Phi}^T + \Phi_V^2 S_{2121}(L-1)\bar{\Phi}^T + \Phi_V S_{2131}(L-1)\bar{\Phi}^T] \\
&-\bar{p}_{2,2}(L)[\bar{H}\bar{\Phi}S_{3111}(L-1)\bar{\Phi}^T + \Phi_V^2 S_{3121}(L-1)\bar{\Phi}^T + \Phi_V S_{3131}(L-1)\bar{\Phi}^T]\} \ , \quad S_{1121}(0) = 0
\end{aligned} \tag{3.46}
$$

$$
\begin{aligned}
&S_{1122}(L) = \tilde{J}_{12}(L,L)[\ p(L)I_m \quad 0\ ]\Phi_V K_V(L,L) + \Phi_V S_{1122}(L-1)\Phi_V^T \\
&-\tilde{J}_{12}(L,L)\{\bar{p}_{2,2}(L)[\bar{H}\bar{\Phi}S_{1112}(L-1)\Phi_V^T + \Phi_V^2 S_{1122}(L-1)\Phi_V^T + \Phi_V S_{1132}(L-1)\Phi_V^T] \\
&+\bar{p}(L)[\bar{H}\bar{\Phi}S_{2112}(L-1)\Phi_V^T + \Phi_V^2 S_{2122}(L-1)\Phi_V^T + \Phi_V S_{2132}(L-1)\Phi_V^T] \\
&-\bar{p}_{2,2}(L)[\bar{H}\bar{\Phi}S_{3112}(L-1)\Phi_V^T + \Phi_V^2 S_{3122}(L-1)\Phi_V^T + \Phi_V S_{3132}(L-1)\Phi_V^T]\} \ , \quad S_{1122}(0) = 0
\end{aligned} \tag{3.47}
$$

$$
\begin{aligned}
&S_{1123}(L) = \tilde{J}_{12}(L,L)[\ p(L)I_m \quad 0\ ]R_U(L) + \Phi_V S_{1123}(L-1)\Phi_V^T \\
&-\tilde{J}_{12}(L,L)\{\bar{p}_{2,2}(L)[\bar{H}\bar{\Phi}S_{1113}(L-1)\Phi_V^T + \Phi_V^2 S_{1123}(L-1)\Phi_V^T + \Phi_V S_{1133}(L-1)\Phi_V^T] \\
&+\bar{p}(L)[\bar{H}\bar{\Phi}S_{2113}(L-1)\Phi_V^T + \Phi_V^2 S_{2123}(L-1)\Phi_V^T + \Phi_V S_{2133}(L-1)\Phi_V^T] \\
&-\bar{p}_{2,2}(L)[\bar{H}\bar{\Phi}S_{3113}(L-1)\Phi_V^T + \Phi_V^2 S_{3123}(L-1)\Phi_V^T + \Phi_V S_{3133}(L-1)\Phi_V^T]\} \ , \quad S_{1123}(0) = 0
\end{aligned} \tag{3.48}
$$

$$
\begin{aligned}
&S_{1131}(L) = \tilde{J}_{13}(L,L)[\ p(L)I_m \quad 0\ ]\bar{H}K_{\bar{x}}(L,L) + \Phi_V S_{1131}(L-1)\bar{\Phi}^T \\
&-\tilde{J}_{13}(L,L)\{\bar{p}_{2,2}(L)[\bar{H}\bar{\Phi}S_{1111}(L-1)\bar{\Phi}^T + \Phi_V^2 S_{1121}(L-1)\bar{\Phi}^T + \Phi_V S_{1131}(L-1)\bar{\Phi}^T] \\
&+\bar{p}(L)[\bar{H}\bar{\Phi}S_{2111}(L-1)\bar{\Phi}^T + \Phi_V^2 S_{2121}(L-1)\bar{\Phi}^T + \Phi_V S_{2131}(L-1)\bar{\Phi}^T] \\
&-\bar{p}_{2,2}(L)[\bar{H}\bar{\Phi}S_{3111}(L-1)\bar{\Phi}^T + \Phi_V^2 S_{3121}(L-1)\bar{\Phi}^T + \Phi_V S_{3131}(L-1)\bar{\Phi}^T]\} \ , \quad S_{1131}(0) = 0
\end{aligned} \tag{3.49}
$$

$$
\begin{aligned}
&S_{1132}(L) = \tilde{J}_{13}(L,L)[\ p(L)I_m \quad 0\ ]\Phi_V K_V(L,L) + \Phi_V S_{1132}(L-1)\Phi_V^T \\
&-\tilde{J}_{13}(L,L)\{\bar{p}_{2,2}(L)[\bar{H}\bar{\Phi}S_{1112}(L-1)\Phi_V^T + \Phi_V^2 S_{1122}(L-1)\Phi_V^T + \Phi_V S_{1132}(L-1)\Phi_V^T] \\
&+\bar{p}(L)[\bar{H}\bar{\Phi}S_{2112}(L-1)\Phi_V^T + \Phi_V^2 S_{2122}(L-1)\Phi_V^T + \Phi_V S_{2132}(L-1)\Phi_V^T] \\
&-\bar{p}_{2,2}(L)[\bar{H}\bar{\Phi}S_{3112}(L-1)\Phi_V^T + \Phi_V^2 S_{3122}(L-1)\Phi_V^T + \Phi_V S_{3132}(L-1)\Phi_V^T]\} \ , \quad S_{1132}(0) = 0
\end{aligned} \tag{3.50}
$$

$$
\begin{aligned}
&S_{1133}(L) = \tilde{J}_{13}(L,L)[\ p(L)I_m \quad 0\ ]R_U(L) + \Phi_V S_{1133}(L-1)\Phi_V^T \\
&-\tilde{J}_{13}(L,L)\{\bar{p}_{2,2}(L)[\bar{H}\bar{\Phi}S_{1113}(L-1)\Phi_V^T + \Phi_V^2 S_{1123}(L-1)\Phi_V^T + \Phi_V S_{1133}(L-1)\Phi_V^T] \\
&+\bar{p}(L)[\bar{H}\bar{\Phi}S_{2113}(L-1)\Phi_V^T + \Phi_V^2 S_{2123}(L-1)\Phi_V^T + \Phi_V S_{2133}(L-1)\Phi_V^T] \\
&-\bar{p}_{2,2}(L)[\bar{H}\bar{\Phi}S_{3113}(L-1)\Phi_V^T + \Phi_V^2 S_{3123}(L-1)\Phi_V^T + \Phi_V S_{3133}(L-1)\Phi_V^T]\} \ , \quad S_{1133}(0) = 0
\end{aligned} \tag{3.51}
$$

$$S_{1211}(L) = \tilde{J}_{11}(L,L)[\ 0 \quad I_m\ ]\bar{H}K_{\bar{x}}(L,L) + \bar{\Phi}S_{1211}(L-1)\bar{\Phi}^T$$
$$-\tilde{J}_{11}(L,L)\{\bar{p}_{2,2}(L)[\bar{H}\bar{\Phi}S_{1211}(L-1)\bar{\Phi}^T + \Phi_V^2 S_{1221}(L-1)\bar{\Phi}^T + \Phi_V S_{1231}(L-1)\bar{\Phi}^T]$$
$$+\bar{p}(L)[\bar{H}\bar{\Phi}S_{2211}(L-1)\bar{\Phi}^T + \Phi_V^2 S_{2221}(L-1)\bar{\Phi}^T + \Phi_V S_{2231}(L-1)\bar{\Phi}^T]$$
$$-\bar{p}_{2,2}(L)[\bar{H}\bar{\Phi}S_{3211}(L-1)\bar{\Phi}^T + \Phi_V^2 S_{3221}(L-1)\bar{\Phi}^T + \Phi_V S_{3231}(L-1)\bar{\Phi}^T]\}, \quad S_{1211}(0) = 0 \tag{3.52}$$

$$S_{1212}(L) = \tilde{J}_{11}(L,L)[\ 0 \quad I_m\ ]\Phi_V K_V(L,L) + \bar{\Phi}S_{1212}(L-1)\Phi_V^T$$
$$-\tilde{J}_{11}(L,L)\{\bar{p}_{2,2}(L)[\bar{H}\bar{\Phi}S_{1212}(L-1)\Phi_V^T + \Phi_V^2 S_{1222}(L-1)\Phi_V^T + \Phi_V S_{1232}(L-1)\Phi_V^T]$$
$$+\bar{p}(L)[\bar{H}\bar{\Phi}S_{2212}(L-1)\Phi_V^T + \Phi_V^2 S_{2222}(L-1)\Phi_V^T + \Phi_V S_{2232}(L-1)\Phi_V^T]$$
$$-\bar{p}_{2,2}(L)[\bar{H}\bar{\Phi}S_{3212}(L-1)\Phi_V^T + \Phi_V^2 S_{3222}(L-1)\Phi_V^T + \Phi_V S_{3232}(L-1)\Phi_V^T]\}, \quad S_{1212}(0) = 0 \tag{3.53}$$

$$S_{1213}(L) = \tilde{J}_{11}(L,L)[\ 0 \quad I_m\ ]R_U(L) + \bar{\Phi}S_{1213}(L-1)\Phi_V^T$$
$$-\tilde{J}_{11}(L,L)\{\bar{p}_{2,2}(L)[\bar{H}\bar{\Phi}S_{1213}(L-1)\Phi_V^T + \Phi_V^2 S_{1223}(L-1)\Phi_V^T + \Phi_V S_{1233}(L-1)\Phi_V^T]$$
$$+\bar{p}(L)[\bar{H}\bar{\Phi}S_{2213}(L-1)\Phi_V^T + \Phi_V^2 S_{2223}(L-1)\Phi_V^T + \Phi_V S_{2233}(L-1)\Phi_V^T]$$
$$-\bar{p}_{2,2}(L)[\bar{H}\bar{\Phi}S_{3213}(L-1)\Phi_V^T + \Phi_V^2 S_{3223}(L-1)\Phi_V^T + \Phi_V S_{3233}(L-1)\Phi_V^T]\}, \quad S_{1213}(0) = 0 \tag{3.54}$$

$$S_{1221}(L) = \tilde{J}_{12}(L,L)[\ 0 \quad I_m\ ]\bar{H}K_{\bar{x}}(L,L) + \Phi_V S_{1221}(L-1)\bar{\Phi}^T$$
$$-\tilde{J}_{12}(L,L)\{\bar{p}_{2,2}(L)[\bar{H}\bar{\Phi}S_{1211}(L-1)\bar{\Phi}^T + \Phi_V^2 S_{1221}(L-1)\bar{\Phi}^T + \Phi_V S_{1231}(L-1)\bar{\Phi}^T]$$
$$+\bar{p}(L)[\bar{H}\bar{\Phi}S_{2211}(L-1)\bar{\Phi}^T + \Phi_V^2 S_{2221}(L-1)\bar{\Phi}^T + \Phi_V S_{2231}(L-1)\bar{\Phi}^T]$$
$$-\bar{p}_{2,2}(L)[\bar{H}\bar{\Phi}S_{3211}(L-1)\bar{\Phi}^T + \Phi_V^2 S_{3221}(L-1)\bar{\Phi}^T + \Phi_V S_{3231}(L-1)\bar{\Phi}^T]\}, \quad S_{1221}(0) = 0 \tag{3.55}$$

$$S_{1222}(L) = \tilde{J}_{12}(L,L)[\ 0 \quad I_m\ ]\Phi_V K_V(L,L) + \Phi_V S_{1222}(L-1)\Phi_V^T$$
$$-\tilde{J}_{12}(L,L)\{\bar{p}_{2,2}(L)[\bar{H}\bar{\Phi}S_{1212}(L-1)\Phi_V^T + \Phi_V^2 S_{1222}(L-1)\Phi_V^T + \Phi_V S_{1232}(L-1)\Phi_V^T]$$
$$+\bar{p}(L)[\bar{H}\bar{\Phi}S_{2212}(L-1)\Phi_V^T + \Phi_V^2 S_{2222}(L-1)\Phi_V^T + \Phi_V S_{2232}(L-1)\Phi_V^T]$$
$$-\bar{p}_{2,2}(L)[\bar{H}\bar{\Phi}S_{3212}(L-1)\Phi_V^T + \Phi_V^2 S_{3222}(L-1)\Phi_V^T + \Phi_V S_{3232}(L-1)\Phi_V^T]\}, \quad S_{1222}(0) = 0 \tag{3.56}$$

$$S_{1223}(L) = \tilde{J}_{12}(L,L)[\ 0 \quad I_m\ ]R_U(L) + \Phi_V S_{1223}(L-1)\Phi_V^T$$
$$-\tilde{J}_{12}(L,L)\{\bar{p}_{2,2}(L)[\bar{H}\bar{\Phi}S_{1213}(L-1)\Phi_V^T + \Phi_V^2 S_{1223}(L-1)\Phi_V^T + \Phi_V S_{1233}(L-1)\Phi_V^T]$$
$$+\bar{p}(L)[\bar{H}\bar{\Phi}S_{2213}(L-1)\Phi_V^T + \Phi_V^2 S_{2223}(L-1)\Phi_V^T + \Phi_V S_{2233}(L-1)\Phi_V^T]$$
$$-\bar{p}_{2,2}(L)[\bar{H}\bar{\Phi}S_{3213}(L-1)\Phi_V^T + \Phi_V^2 S_{3223}(L-1)\Phi_V^T + \Phi_V S_{3233}(L-1)\Phi_V^T]\}, \quad S_{1223}(0) = 0 \tag{3.57}$$

$$S_{1231}(L) = \tilde{J}_{13}(L,L)[\ 0 \quad I_m\ ]\bar{H}K_{\bar{x}}(L,L) + \Phi_V S_{1231}(L-1)\bar{\Phi}^T$$
$$-\tilde{J}_{13}(L,L)\{\bar{p}_{2,2}(L)[\bar{H}\bar{\Phi}S_{1211}(L-1)\bar{\Phi}^T + \Phi_V^2 S_{1221}(L-1)\bar{\Phi}^T + \Phi_V S_{1231}(L-1)\bar{\Phi}^T]$$
$$+\bar{p}(L)[\bar{H}\bar{\Phi}S_{2211}(L-1)\bar{\Phi}^T + \Phi_V^2 S_{2221}(L-1)\bar{\Phi}^T + \Phi_V S_{2231}(L-1)\bar{\Phi}^T]$$
$$-\bar{p}_{2,2}(L)[\bar{H}\bar{\Phi}S_{3211}(L-1)\bar{\Phi}^T + \Phi_V^2 S_{3221}(L-1)\bar{\Phi}^T + \Phi_V S_{3231}(L-1)\bar{\Phi}^T]\}, \quad S_{1231}(0) = 0 \tag{3.58}$$

$$
\begin{aligned}
S_{1232}(L) &= \tilde{J}_{13}(L,L)[\ 0 \quad I_m\ ]\Phi_V K_V(L,L) + \Phi_V S_{1232}(L-1)\Phi_V^T \\
&- \tilde{J}_{13}(L,L)\{\bar{p}_{2,2}(L)[\bar{H}\bar{\Phi}S_{1212}(L-1)\Phi_V^T + \Phi_V^2 S_{1222}(L-1)\Phi_V^T + \Phi_V S_{1232}(L-1)\Phi_V^T] \\
&+ \bar{p}(L)[\bar{H}\bar{\Phi}S_{2212}(L-1)\Phi_V^T + \Phi_V^2 S_{2222}(L-1)\Phi_V^T + \Phi_V S_{2232}(L-1)\Phi_V^T] \\
&- \bar{p}_{2,2}(L)[\bar{H}\bar{\Phi}S_{3212}(L-1)\Phi_V^T + \Phi_V^2 S_{3222}(L-1)\Phi_V^T + \Phi_V S_{3232}(L-1)\Phi_V^T]\}\ , \quad S_{1232}(0) = 0
\end{aligned}
\tag{3.59}
$$

$$
\begin{aligned}
S_{1233}(L) &= \tilde{J}_{13}(L,L)[\ 0 \quad I_m\ ]R_U(L) + \Phi_V S_{1233}(L-1)\Phi_V^T \\
&- \tilde{J}_{13}(L,L)\{\bar{p}_{2,2}(L)[\bar{H}\bar{\Phi}S_{1213}(L-1)\Phi_V^T + \Phi_V^2 S_{1223}(L-1)\Phi_V^T + \Phi_V S_{1233}(L-1)\Phi_V^T] \\
&+ \bar{p}(L)[\bar{H}\bar{\Phi}S_{2213}(L-1)\Phi_V^T + \Phi_V^2 S_{2223}(L-1)\Phi_V^T + \Phi_V S_{2233}(L-1)\Phi_V^T] \\
&- \bar{p}_{2,2}(L)[\bar{H}\bar{\Phi}S_{3213}(L-1)\Phi_V^T + \Phi_V^2 S_{3223}(L-1)\Phi_V^T + \Phi_V S_{3233}(L-1)\Phi_V^T]\}\ , \quad S_{1233}(0) = 0
\end{aligned}
\tag{3.60}
$$

$$
\begin{aligned}
S_{1311}(L) &= \tilde{J}_{11}(L,L)[\ 0 \quad p(L)I_m\ ]\bar{H}K_{\bar{x}}(L,L) + \bar{\Phi}S_{1311}(L-1)\bar{\Phi}^T \\
&- \tilde{J}_{11}(L,L)\{\bar{p}_{2,2}(L)[\bar{H}\bar{\Phi}S_{1311}(L-1)\bar{\Phi}^T + \Phi_V^2 S_{1321}(L-1)\bar{\Phi}^T + \Phi_V S_{1331}(L-1)\bar{\Phi}^T] \\
&+ \bar{p}(L)[\bar{H}\bar{\Phi}S_{2311}(L-1)\bar{\Phi}^T + \Phi_V^2 S_{2321}(L-1)\bar{\Phi}^T + \Phi_V S_{2331}(L-1)\bar{\Phi}^T] \\
&- \bar{p}_{2,2}(L)[\bar{H}\bar{\Phi}S_{3311}(L-1)\bar{\Phi}^T + \Phi_V^2 S_{3321}(L-1)\bar{\Phi}^T + \Phi_V S_{3331}(L-1)\bar{\Phi}^T]\}\ , \quad S_{1311}(0) = 0
\end{aligned}
\tag{3.61}
$$

$$
\begin{aligned}
S_{1312}(L) &= \tilde{J}_{11}(L,L)[\ 0 \quad p(L)I_m\ ]\Phi_V K_V(L,L) + \bar{\Phi}S_{1312}(L-1)\Phi_V^T \\
&- \tilde{J}_{11}(L,L)\{\bar{p}_{2,2}(L)[\bar{H}\bar{\Phi}S_{1312}(L-1)\Phi_V^T + \Phi_V^2 S_{1322}(L-1)\Phi_V^T + \Phi_V S_{1332}(L-1)\Phi_V^T] \\
&+ \bar{p}(L)[\bar{H}\bar{\Phi}S_{2312}(L-1)\Phi_V^T + \Phi_V^2 S_{2322}(L-1)\Phi_V^T + \Phi_V S_{2332}(L-1)\Phi_V^T] \\
&- \bar{p}_{2,2}(L)[\bar{H}\bar{\Phi}S_{3312}(L-1)\Phi_V^T + \Phi_V^2 S_{3322}(L-1)\Phi_V^T + \Phi_V S_{3332}(L-1)\Phi_V^T]\}\ , \quad S_{1312}(0) = 0
\end{aligned}
\tag{3.62}
$$

$$
\begin{aligned}
S_{1313}(L) &= \tilde{J}_{11}(L,L)[\ 0 \quad p(L)I_m\ ]R_U(L) + \bar{\Phi}S_{1313}(L-1)\Phi_V^T \\
&- \tilde{J}_{11}(L,L)\{\bar{p}_{2,2}(L)[\bar{H}\bar{\Phi}S_{1313}(L-1)\Phi_V^T + \Phi_V^2 S_{1323}(L-1)\Phi_V^T + \Phi_V S_{1333}(L-1)\Phi_V^T] \\
&+ \bar{p}(L)[\bar{H}\bar{\Phi}S_{2313}(L-1)\Phi_V^T + \Phi_V^2 S_{2323}(L-1)\Phi_V^T + \Phi_V S_{2333}(L-1)\Phi_V^T] \\
&- \bar{p}_{2,2}(L)[\bar{H}\bar{\Phi}S_{3313}(L-1)\Phi_V^T + \Phi_V^2 S_{3323}(L-1)\Phi_V^T + \Phi_V S_{3333}(L-1)\Phi_V^T]\}\ , \quad S_{1313}(0) = 0
\end{aligned}
\tag{3.63}
$$

$$
\begin{aligned}
S_{1321}(L) &= \tilde{J}_{12}(L,L)[\ 0 \quad p(L)I_m\ ]\bar{H}K_{\bar{x}}(L,L) + \Phi_V S_{1321}(L-1)\bar{\Phi}^T \\
&- \tilde{J}_{12}(L,L)\{\bar{p}_{2,2}(L)[\bar{H}\bar{\Phi}S_{1311}(L-1)\bar{\Phi}^T + \Phi_V^2 S_{1321}(L-1)\bar{\Phi}^T + \Phi_V S_{1331}(L-1)\bar{\Phi}^T] \\
&+ \bar{p}(L)[\bar{H}\bar{\Phi}S_{2311}(L-1)\bar{\Phi}^T + \Phi_V^2 S_{2321}(L-1)\bar{\Phi}^T + \Phi_V S_{2331}(L-1)\bar{\Phi}^T] \\
&- \bar{p}_{2,2}(L)[\bar{H}\bar{\Phi}S_{3311}(L-1)\bar{\Phi}^T + \Phi_V^2 S_{3321}(L-1)\bar{\Phi}^T + \Phi_V S_{3331}(L-1)\bar{\Phi}^T]\}\ , \quad S_{1321}(0) = 0
\end{aligned}
\tag{3.64}
$$

$$\begin{aligned}
S_{1322}(L) &= \tilde{J}_{12}(L,L)[\ 0 \quad p(L)I_m\ ]\Phi_V K_V(L,L) + \Phi_V S_{1322}(L-1)\Phi_V^T \\
&-\tilde{J}_{12}(L,L)\{\bar{p}_{2,2}(L)[\bar{H}\bar{\Phi}S_{1312}(L-1)\Phi_V^T + \Phi_V^2 S_{1322}(L-1)\Phi_V^T + \Phi_V S_{1332}(L-1)\Phi_V^T] \\
&+\bar{p}(L)[\bar{H}\bar{\Phi}S_{2312}(L-1)\Phi_V^T + \Phi_V^2 S_{2322}(L-1)\Phi_V^T + \Phi_V S_{2332}(L-1)\Phi_V^T] \\
&-\bar{p}_{2,2}(L)[\bar{H}\bar{\Phi}S_{3312}(L-1)\Phi_V^T + \Phi_V^2 S_{3322}(L-1)\Phi_V^T + \Phi_V S_{3332}(L-1)\Phi_V^T]\}\ , \quad S_{1322}(0)=0
\end{aligned} \tag{3.65}$$

$$\begin{aligned}
S_{1323}(L) &= \tilde{J}_{12}(L,L)[\ 0 \quad p(L)I_m\ ]R_U(L) + \Phi_V S_{1323}(L-1)\Phi_V^T \\
&-\tilde{J}_{12}(L,L)\{\bar{p}_{2,2}(L)[\bar{H}\bar{\Phi}S_{1313}(L-1)\Phi_V^T + \Phi_V^2 S_{1323}(L-1)\Phi_V^T + \Phi_V S_{1333}(L-1)\Phi_V^T] \\
&+\bar{p}(L)[\bar{H}\bar{\Phi}S_{2313}(L-1)\Phi_V^T + \Phi_V^2 S_{2323}(L-1)\Phi_V^T + \Phi_V S_{2333}(L-1)\Phi_V^T] \\
&-\bar{p}_{2,2}(L)[\bar{H}\bar{\Phi}S_{3313}(L-1)\Phi_V^T + \Phi_V^2 S_{3323}(L-1)\Phi_V^T + \Phi_V S_{3333}(L-1)\Phi_V^T]\}\ , \quad S_{1323}(0)=0
\end{aligned} \tag{3.66}$$

$$\begin{aligned}
S_{1331}(L) &= \tilde{J}_{13}(L,L)[\ 0 \quad p(L)I_m\ ]\bar{H}K_{\bar{x}}(L,L) + \Phi_V S_{1331}(L-1)\bar{\Phi}^T \\
&-\tilde{J}_{13}(L,L)\{\bar{p}_{2,2}(L)[\bar{H}\bar{\Phi}S_{1311}(L-1)\bar{\Phi}^T + \Phi_V^2 S_{1321}(L-1)\bar{\Phi}^T + \Phi_V S_{1331}(L-1)\bar{\Phi}^T] \\
&+\bar{p}(L)[\bar{H}\bar{\Phi}S_{2311}(L-1)\bar{\Phi}^T + \Phi_V^2 S_{2321}(L-1)\bar{\Phi}^T + \Phi_V S_{2331}(L-1)\bar{\Phi}^T] \\
&-\bar{p}_{2,2}(L)[\bar{H}\bar{\Phi}S_{3311}(L-1)\bar{\Phi}^T + \Phi_V^2 S_{3321}(L-1)\bar{\Phi}^T + \Phi_V S_{3331}(L-1)\bar{\Phi}^T]\}\ , \quad S_{1331}(0)=0
\end{aligned} \tag{3.67}$$

$$\begin{aligned}
S_{1332}(L) &= \tilde{J}_{13}(L,L)[\ 0 \quad p(L)I_m\ ]\Phi_V K_V(L,L) + \Phi_V S_{1332}(L-1)\Phi_V^T \\
&-\tilde{J}_{13}(L,L)\{\bar{p}_{2,2}(L)[\bar{H}\bar{\Phi}S_{1312}(L-1)\Phi_V^T + \Phi_V^2 S_{1322}(L-1)\Phi_V^T + \Phi_V S_{1332}(L-1)\Phi_V^T] \\
&+\bar{p}(L)[\bar{H}\bar{\Phi}S_{2312}(L-1)\Phi_V^T + \Phi_V^2 S_{2322}(L-1)\Phi_V^T + \Phi_V S_{2332}(L-1)\Phi_V^T] \\
&-\bar{p}_{2,2}(L)[\bar{H}\bar{\Phi}S_{3312}(L-1)\Phi_V^T + \Phi_V^2 S_{3322}(L-1)\Phi_V^T + \Phi_V S_{3332}(L-1)\Phi_V^T]\}\ , \quad S_{1332}(0)=0
\end{aligned} \tag{3.68}$$

$$\begin{aligned}
S_{1333}(L) &= \tilde{J}_{13}(L,L)[\ 0 \quad p(L)I_m\ ]R_U(L) + \Phi_V S_{1333}(L-1)\Phi_V^T \\
&-\tilde{J}_{13}(L,L)\{\bar{p}_{2,2}(L)[\bar{H}\bar{\Phi}S_{1313}(L-1)\Phi_V^T + \Phi_V^2 S_{1323}(L-1)\Phi_V^T + \Phi_V S_{1333}(L-1)\Phi_V^T] \\
&+\bar{p}(L)[\bar{H}\bar{\Phi}S_{2313}(L-1)\Phi_V^T + \Phi_V^2 S_{2323}(L-1)\Phi_V^T + \Phi_V S_{2333}(L-1)\Phi_V^T] \\
&-\bar{p}_{2,2}(L)[\bar{H}\bar{\Phi}S_{3313}(L-1)\Phi_V^T + \Phi_V^2 S_{3323}(L-1)\Phi_V^T + \Phi_V S_{3333}(L-1)\Phi_V^T]\}\ , \quad S_{1333}(0)=0
\end{aligned} \tag{3.69}$$

$$\begin{aligned}
S_{2111}(L) &= \tilde{J}_{21}(L,L)[\ p(L)I_m \quad 0\ ]\bar{H}K_{\bar{x}}(L,L) + \bar{\Phi}S_{2111}(L-1)\bar{\Phi}^T \\
&-\tilde{J}_{21}(L,L)\{\bar{p}_{2,2}(L)[\bar{H}\bar{\Phi}S_{1111}(L-1)\bar{\Phi}^T + \Phi_V^2 S_{1121}(L-1)\bar{\Phi}^T + \Phi_V S_{1131}(L-1)\bar{\Phi}^T] \\
&+\bar{p}(L)[\bar{H}\bar{\Phi}S_{2111}(L-1)\bar{\Phi}^T + \Phi_V^2 S_{2121}(L-1)\bar{\Phi}^T + \Phi_V S_{2131}(L-1)\bar{\Phi}^T] \\
&-\bar{p}_{2,2}(L)[\bar{H}\bar{\Phi}S_{3111}(L-1)\bar{\Phi}^T + \Phi_V^2 S_{3121}(L-1)\bar{\Phi}^T + \Phi_V S_{3131}(L-1)\bar{\Phi}^T]\}\ , \quad S_{2111}(0)=0
\end{aligned} \tag{3.70}$$

$$\begin{aligned}
S_{2112}(L) &= \tilde{J}_{21}(L,L)[\ p(L)I_m \quad 0\ ]\Phi_V K_V(L,L) + \bar{\Phi}S_{2112}(L-1)\Phi_V^T \\
&-\tilde{J}_{21}(L,L)\{\bar{p}_{2,2}(L)[\bar{H}\bar{\Phi}S_{1112}(L-1)\Phi_V^T + \Phi_V^2 S_{1122}(L-1)\Phi_V^T + \Phi_V S_{1132}(L-1)\Phi_V^T] \\
&+\bar{p}(L)[\bar{H}\bar{\Phi}S_{2112}(L-1)\Phi_V^T + \Phi_V^2 S_{2122}(L-1)\Phi_V^T + \Phi_V S_{2132}(L-1)\Phi_V^T] \\
&-\bar{p}_{2,2}(L)[\bar{H}\bar{\Phi}S_{3112}(L-1)\Phi_V^T + \Phi_V^2 S_{3122}(L-1)\Phi_V^T + \Phi_V S_{3132}(L-1)\Phi_V^T]\}\ , \quad S_{2112}(0)=0
\end{aligned} \tag{3.71}$$

$$\begin{aligned} S_{2113}(L) &= \tilde{J}_{21}(L,L)[\ p(L)I_m \quad 0\ ]R_U(L) + \bar{\Phi} S_{2113}(L-1)\Phi_V{}^T \\ &-\tilde{J}_{21}(L,L)\{\bar{p}_{2,2}(L)[\bar{H}\bar{\Phi}S_{1113}(L-1)\Phi_V^T + \Phi_V^2 S_{1123}(L-1)\Phi_V^T + \Phi_V S_{1133}(L-1)\Phi_V^T] \\ &+\bar{p}(L)[\bar{H}\bar{\Phi}S_{2113}(L-1)\Phi_V^T + \Phi_V^2 S_{2123}(L-1)\Phi_V^T + \Phi_V S_{2133}(L-1)\Phi_V^T] \\ &-\bar{p}_{2,2}(L)[\bar{H}\bar{\Phi}S_{3113}(L-1)\Phi_V^T + \Phi_V^2 S_{3123}(L-1)\Phi_V^T + \Phi_V S_{3133}(L-1)\Phi_V^T]\}\ , \quad S_{2113}(0)=0 \end{aligned} \tag{3.72}$$

$$\begin{aligned} S_{2121}(L) &= \tilde{J}_{22}(L,L)[\ p(L)I_m \quad 0\ ]\bar{H}K_{\bar{x}}(L,L) + \Phi_V S_{2121}(L-1)\bar{\Phi}^T \\ &-\tilde{J}_{22}(L,L)\{\bar{p}_{2,2}(L)[\bar{H}\bar{\Phi}S_{1111}(L-1)\bar{\Phi}^T + \Phi_V^2 S_{1121}(L-1)\bar{\Phi}^T + \Phi_V S_{1131}(L-1)\bar{\Phi}^T] \\ &+\bar{p}(L)[\bar{H}\bar{\Phi}S_{2111}(L-1)\bar{\Phi}^T + \Phi_V^2 S_{2121}(L-1)\bar{\Phi}^T + \Phi_V S_{2131}(L-1)\bar{\Phi}^T] \\ &-\bar{p}_{2,2}(L)[\bar{H}\bar{\Phi}S_{3111}(L-1)\bar{\Phi}^T + \Phi_V^2 S_{3121}(L-1)\bar{\Phi}^T + \Phi_V S_{3131}(L-1)\bar{\Phi}^T]\}\ , \quad S_{2121}(0)=0 \end{aligned} \tag{3.73}$$

$$\begin{aligned} S_{2122}(L) &= \tilde{J}_{22}(L,L)[\ p(L)I_m \quad 0\ ]\Phi_V K_V(L,L) + \Phi_V S_{2122}(L-1)\Phi_V^T \\ &-\tilde{J}_{22}(L,L)\{\bar{p}_{2,2}(L)[\bar{H}\bar{\Phi}S_{1112}(L-1)\Phi_V^T + \Phi_V^2 S_{1122}(L-1)\Phi_V^T + \Phi_V S_{1132}(L-1)\Phi_V^T] \\ &+\bar{p}(L)[\bar{H}\bar{\Phi}S_{2112}(L-1)\Phi_V^T + \Phi_V^2 S_{2122}(L-1)\Phi_V^T + \Phi_V S_{2132}(L-1)\Phi_V^T] \\ &-\bar{p}_{2,2}(L)[\bar{H}\bar{\Phi}S_{3112}(L-1)\Phi_V^T + \Phi_V^2 S_{3122}(L-1)\Phi_V^T + \Phi_V S_{3132}(L-1)\Phi_V^T]\}\ , \quad S_{2122}(0)=0 \end{aligned} \tag{3.74}$$

$$\begin{aligned} S_{2123}(L) &= \tilde{J}_{22}(L,L)[\ p(L)I_m \quad 0\ ]R_U(L) + \Phi_V S_{2123}(L-1)\Phi_V^T \\ &-\tilde{J}_{22}(L,L)\{\bar{p}_{2,2}(L)[\bar{H}\bar{\Phi}S_{1113}(L-1)\Phi_V^T + \Phi_V^2 S_{1123}(L-1)\Phi_V^T + \Phi_V S_{1133}(L-1)\Phi_V^T] \\ &+\bar{p}(L)[\bar{H}\bar{\Phi}S_{2113}(L-1)\Phi_V^T + \Phi_V^2 S_{2123}(L-1)\Phi_V^T + \Phi_V S_{2133}(L-1)\Phi_V^T] \\ &-\bar{p}_{2,2}(L)[\bar{H}\bar{\Phi}S_{3113}(L-1)\Phi_V^T + \Phi_V^2 S_{3123}(L-1)\Phi_V^T + \Phi_V S_{3133}(L-1)\Phi_V^T]\}\ , \quad S_{2123}(0)=0 \end{aligned} \tag{3.75}$$

$$\begin{aligned} S_{2131}(L) &= \tilde{J}_{23}(L,L)[\ p(L)I_m \quad 0\ ]\bar{H}K_{\bar{x}}(L,L) + \Phi_V S_{2131}(L-1)\bar{\Phi}^T \\ &-\tilde{J}_{23}(L,L)\{\bar{p}_{2,2}(L)[\bar{H}\bar{\Phi}S_{1111}(L-1)\bar{\Phi}^T + \Phi_V^2 S_{1121}(L-1)\bar{\Phi}^T + \Phi_V S_{1131}(L-1)\bar{\Phi}^T] \\ &+\bar{p}(L)[\bar{H}\bar{\Phi}S_{2111}(L-1)\bar{\Phi}^T + \Phi_V^2 S_{2121}(L-1)\bar{\Phi}^T + \Phi_V S_{2131}(L-1)\bar{\Phi}^T] \\ &-\bar{p}_{2,2}(L)[\bar{H}\bar{\Phi}S_{3111}(L-1)\bar{\Phi}^T + \Phi_V^2 S_{3121}(L-1)\bar{\Phi}^T + \Phi_V S_{3131}(L-1)\bar{\Phi}^T]\}\ , \quad S_{2131}(0)=0 \end{aligned} \tag{3.76}$$

$$\begin{aligned} S_{2132}(L) &= \tilde{J}_{23}(L,L)[\ p(L)I_m \quad 0\ ]\Phi_V K_V(L,L) + \Phi_V S_{2132}(L-1)\Phi_V^T \\ &-\tilde{J}_{23}(L,L)\{\bar{p}_{2,2}(L)[\bar{H}\bar{\Phi}S_{1112}(L-1)\Phi_V^T + \Phi_V^2 S_{1122}(L-1)\Phi_V^T + \Phi_V S_{1132}(L-1)\Phi_V^T] \\ &+\bar{p}(L)[\bar{H}\bar{\Phi}S_{2112}(L-1)\Phi_V^T + \Phi_V^2 S_{2122}(L-1)\Phi_V^T + \Phi_V S_{2132}(L-1)\Phi_V^T] \\ &-\bar{p}_{2,2}(L)[\bar{H}\bar{\Phi}S_{3112}(L-1)\Phi_V^T + \Phi_V^2 S_{3122}(L-1)\Phi_V^T + \Phi_V S_{3132}(L-1)\Phi_V^T]\}\ , \quad S_{2132}(0)=0 \end{aligned} \tag{3.77}$$

$$\begin{aligned} S_{2133}(L) &= \tilde{J}_{23}(L,L)[\ p(L)I_m \quad 0\ ]R_U(L) + \Phi_V S_{2133}(L-1)\Phi_V^T \\ &-\tilde{J}_{23}(L,L)\{\bar{p}_{2,2}(L)[\bar{H}\bar{\Phi}S_{1113}(L-1)\Phi_V^T + \Phi_V^2 S_{1123}(L-1)\Phi_V^T + \Phi_V S_{1133}(L-1)\Phi_V^T] \\ &+\bar{p}(L)[\bar{H}\bar{\Phi}S_{2113}(L-1)\Phi_V^T + \Phi_V^2 S_{2123}(L-1)\Phi_V^T + \Phi_V S_{2133}(L-1)\Phi_V^T] \\ &-\bar{p}_{2,2}(L)[\bar{H}\bar{\Phi}S_{3113}(L-1)\Phi_V^T + \Phi_V^2 S_{3123}(L-1)\Phi_V^T + \Phi_V S_{3133}(L-1)\Phi_V^T]\}\ , \quad S_{2133}(0)=0 \end{aligned} \tag{3.78}$$

$$S_{2211}(L) = \tilde{J}_{21}(L,L)[\ 0 \quad I_m\ ]\bar{H}K_{\bar{x}}(L,L) + \bar{\Phi}S_{2211}(L-1)\bar{\Phi}^T$$
$$-\tilde{J}_{21}(L,L)\{\bar{p}_{2,2}(L)[\bar{H}\bar{\Phi}S_{1211}(L-1)\bar{\Phi}^T + \Phi_V^2 S_{1221}(L-1)\bar{\Phi}^T + \Phi_V S_{1231}(L-1)\bar{\Phi}^T]$$
$$+\bar{p}(L)[\bar{H}\bar{\Phi}S_{2211}(L-1)\bar{\Phi}^T + \Phi_V^2 S_{2221}(L-1)\bar{\Phi}^T + \Phi_V S_{2231}(L-1)\bar{\Phi}^T]$$
$$-\bar{p}_{2,2}(L)[\bar{H}\bar{\Phi}S_{3211}(L-1)\bar{\Phi}^T + \Phi_V^2 S_{3221}(L-1)\bar{\Phi}^T + \Phi_V S_{3231}(L-1)\bar{\Phi}^T]\}\ , \quad S_{2211}(0) = 0 \tag{3.79}$$

$$S_{2212}(L) = \tilde{J}_{21}(L,L)[\ 0 \quad I_m\ ]\Phi_V K_V(L,L) + \bar{\Phi}S_{2212}(L-1)\Phi_V^T$$
$$-\tilde{J}_{21}(L,L)\{\bar{p}_{2,2}(L)[\bar{H}\bar{\Phi}S_{1212}(L-1)\Phi_V^T + \Phi_V^2 S_{1222}(L-1)\Phi_V^T + \Phi_V S_{1232}(L-1)\Phi_V^T]$$
$$+\bar{p}(L)[\bar{H}\bar{\Phi}S_{2212}(L-1)\Phi_V^T + \Phi_V^2 S_{2222}(L-1)\Phi_V^T + \Phi_V S_{2232}(L-1)\Phi_V^T]$$
$$-\bar{p}_{2,2}(L)[\bar{H}\bar{\Phi}S_{3212}(L-1)\Phi_V^T + \Phi_V^2 S_{3222}(L-1)\Phi_V^T + \Phi_V S_{3232}(L-1)\Phi_V^T]\}\ , \quad S_{2212}(0) = 0 \tag{3.80}$$

$$S_{2213}(L) = \tilde{J}_{21}(L,L)[\ 0 \quad I_m\ ]R_U(L) + \bar{\Phi}S_{2213}(L-1)\Phi_V^T$$
$$-\tilde{J}_{21}(L,L)\{\bar{p}_{2,2}(L)[\bar{H}\bar{\Phi}S_{1213}(L-1)\Phi_V^T + \Phi_V^2 S_{1223}(L-1)\Phi_V^T + \Phi_V S_{1233}(L-1)\Phi_V^T]$$
$$+\bar{p}(L)[\bar{H}\bar{\Phi}S_{2213}(L-1)\Phi_V^T + \Phi_V^2 S_{2223}(L-1)\Phi_V^T + \Phi_V S_{2233}(L-1)\Phi_V^T]$$
$$-\bar{p}_{2,2}(L)[\bar{H}\bar{\Phi}S_{3213}(L-1)\Phi_V^T + \Phi_V^2 S_{3223}(L-1)\Phi_V^T + \Phi_V S_{3233}(L-1)\Phi_V^T]\}\ , \quad S_{2213}(0) = 0 \tag{3.81}$$

$$S_{2221}(L) = \tilde{J}_{22}(L,L)[\ 0 \quad I_m\ ]\bar{H}K_{\bar{x}}(L,L) + \Phi_V S_{2221}(L-1)\bar{\Phi}^T$$
$$-\tilde{J}_{22}(L,L)\{\bar{p}_{2,2}(L)[\bar{H}\bar{\Phi}S_{1211}(L-1)\bar{\Phi}^T + \Phi_V^2 S_{1221}(L-1)\bar{\Phi}^T + \Phi_V S_{1231}(L-1)\bar{\Phi}^T]$$
$$+\bar{p}(L)[\bar{H}\bar{\Phi}S_{2211}(L-1)\bar{\Phi}^T + \Phi_V^2 S_{2221}(L-1)\bar{\Phi}^T + \Phi_V S_{2231}(L-1)\bar{\Phi}^T]$$
$$-\bar{p}_{2,2}(L)[\bar{H}\bar{\Phi}S_{3211}(L-1)\bar{\Phi}^T + \Phi_V^2 S_{3221}(L-1)\bar{\Phi}^T + \Phi_V S_{3231}(L-1)\bar{\Phi}^T]\}\ , \quad S_{2221}(0) = 0 \tag{3.82}$$

$$S_{2222}(L) = \tilde{J}_{22}(L,L)[\ 0 \quad I_m\ ]\Phi_V K_V(L,L) + \Phi_V S_{2222}(L-1)\Phi_V^T$$
$$-\tilde{J}_{22}(L,L)\{\bar{p}_{2,2}(L)[\bar{H}\bar{\Phi}S_{1212}(L-1)\Phi_V^T + \Phi_V^2 S_{1222}(L-1)\Phi_V^T + \Phi_V S_{1232}(L-1)\Phi_V^T]$$
$$+\bar{p}(L)[\bar{H}\bar{\Phi}S_{2212}(L-1)\Phi_V^T + \Phi_V^2 S_{2222}(L-1)\Phi_V^T + \Phi_V S_{2232}(L-1)\Phi_V^T]$$
$$-\bar{p}_{2,2}(L)[\bar{H}\bar{\Phi}S_{3212}(L-1)\Phi_V^T + \Phi_V^2 S_{3222}(L-1)\Phi_V^T + \Phi_V S_{3232}(L-1)\Phi_V^T]\}\ , \quad S_{2222}(0) = 0 \tag{3.83}$$

$$S_{2223}(L) = \tilde{J}_{22}(L,L)[\ 0 \quad I_m\ ]R_U(L) + \Phi_V S_{2223}(L-1)\Phi_V^T$$
$$-\tilde{J}_{22}(L,L)\{\bar{p}_{2,2}(L)[\bar{H}\bar{\Phi}S_{1213}(L-1)\Phi_V^T + \Phi_V^2 S_{1223}(L-1)\Phi_V^T + \Phi_V S_{1233}(L-1)\Phi_V^T]$$
$$+\bar{p}(L)[\bar{H}\bar{\Phi}S_{2213}(L-1)\Phi_V^T + \Phi_V^2 S_{2223}(L-1)\Phi_V^T + \Phi_V S_{2233}(L-1)\Phi_V^T]$$
$$-\bar{p}_{2,2}(L)[\bar{H}\bar{\Phi}S_{3213}(L-1)\Phi_V^T + \Phi_V^2 S_{3223}(L-1)\Phi_V^T + \Phi_V S_{3233}(L-1)\Phi_V^T]\}\ , \quad S_{2223}(0) = 0 \tag{3.84}$$

$$S_{2231}(L) = \tilde{J}_{23}(L,L)[\ 0 \quad I_m\ ]\bar{H}K_{\bar{x}}(L,L) + \Phi_V S_{2231}(L-1)\bar{\Phi}^T$$
$$-\tilde{J}_{23}(L,L)\{\bar{p}_{2,2}(L)[\bar{H}\bar{\Phi}S_{1211}(L-1)\bar{\Phi}^T + \Phi_V^2 S_{1221}(L-1)\bar{\Phi}^T + \Phi_V S_{1231}(L-1)\bar{\Phi}^T]$$
$$+\bar{p}(L)[\bar{H}\bar{\Phi}S_{2211}(L-1)\bar{\Phi}^T + \Phi_V^2 S_{2221}(L-1)\bar{\Phi}^T + \Phi_V S_{2231}(L-1)\bar{\Phi}^T]$$
$$-\bar{p}_{2,2}(L)[\bar{H}\bar{\Phi}S_{3211}(L-1)\bar{\Phi}^T + \Phi_V^2 S_{3221}(L-1)\bar{\Phi}^T + \Phi_V S_{3231}(L-1)\bar{\Phi}^T]\}\ , \quad S_{2231}(0) = 0 \tag{3.85}$$

$$S_{2232}(L) = \tilde{J}_{23}(L,L)[\ 0 \quad I_m\ ]\Phi_V K_V(L,L) + \Phi_V S_{2232}(L-1)\Phi_V^T$$
$$-\tilde{J}_{23}(L,L)\{\bar{p}_{2,2}(L)[\bar{H}\bar{\Phi}S_{1212}(L-1)\Phi_V^T + \Phi_V^2 S_{1222}(L-1)\Phi_V^T + \Phi_V S_{1232}(L-1)\Phi_V^T]$$
$$+\bar{p}(L)[\bar{H}\bar{\Phi}S_{2212}(L-1)\Phi_V^T + \Phi_V^2 S_{2222}(L-1)\Phi_V^T + \Phi_V S_{2232}(L-1)\Phi_V^T]$$
$$-\bar{p}_{2,2}(L)[\bar{H}\bar{\Phi}S_{3212}(L-1)\Phi_V^T + \Phi_V^2 S_{3222}(L-1)\Phi_V^T + \Phi_V S_{3232}(L-1)\Phi_V^T]\}\ , \quad S_{2232}(0) = 0 \tag{3.86}$$

$$S_{2233}(L) = \tilde{J}_{23}(L,L)[\ 0 \quad I_m\ ]R_U(L) + \Phi_V S_{2233}(L-1)\Phi_V^T$$
$$-\tilde{J}_{23}(L,L)\{\bar{p}_{2,2}(L)[\bar{H}\bar{\Phi}S_{1213}(L-1)\Phi_V^T + \Phi_V^2 S_{1223}(L-1)\Phi_V^T + \Phi_V S_{1233}(L-1)\Phi_V^T]$$
$$+\bar{p}(L)[\bar{H}\bar{\Phi}S_{2213}(L-1)\Phi_V^T + \Phi_V^2 S_{2223}(L-1)\Phi_V^T + \Phi_V S_{2233}(L-1)\Phi_V^T]$$
$$-\bar{p}_{2,2}(L)[\bar{H}\bar{\Phi}S_{3213}(L-1)\Phi_V^T + \Phi_V^2 S_{3223}(L-1)\Phi_V^T + \Phi_V S_{3233}(L-1)\Phi_V^T]\}\ , \quad S_{2233}(0) = 0 \tag{3.87}$$

$$S_{2311}(L) = \tilde{J}_{21}(L,L)[\ 0 \quad p(L)I_m\ ]\bar{H}K_{\bar{x}}(L,L) + \bar{\Phi}S_{2311}(L-1)\bar{\Phi}^T$$
$$-\tilde{J}_{21}(L,L)\{\bar{p}_{2,2}(L)[\bar{H}\bar{\Phi}S_{1311}(L-1)\bar{\Phi}^T + \Phi_V^2 S_{1321}(L-1)\bar{\Phi}^T + \Phi_V S_{1331}(L-1)\bar{\Phi}^T]$$
$$+\bar{p}(L)[\bar{H}\bar{\Phi}S_{2311}(L-1)\bar{\Phi}^T + \Phi_V^2 S_{2321}(L-1)\bar{\Phi}^T + \Phi_V S_{2331}(L-1)\bar{\Phi}^T]$$
$$-\bar{p}_{2,2}(L)[\bar{H}\bar{\Phi}S_{3311}(L-1)\bar{\Phi}^T + \Phi_V^2 S_{3321}(L-1)\bar{\Phi}^T + \Phi_V S_{3331}(L-1)\bar{\Phi}^T]\}\ , \quad S_{2311}(0) = 0 \tag{3.88}$$

$$S_{2312}(L) = \tilde{J}_{21}(L,L)[\ 0 \quad p(L)I_m\ ]\Phi_V K_V(L,L) + \bar{\Phi}S_{2312}(L-1)\Phi_V^T$$
$$-\tilde{J}_{21}(L,L)\{\bar{p}_{2,2}(L)[\bar{H}\bar{\Phi}S_{1312}(L-1)\Phi_V^T + \Phi_V^2 S_{1322}(L-1)\Phi_V^T + \Phi_V S_{1332}(L-1)\Phi_V^T]$$
$$+\bar{p}(L)[\bar{H}\bar{\Phi}S_{2312}(L-1)\Phi_V^T + \Phi_V^2 S_{2322}(L-1)\Phi_V^T + \Phi_V S_{2332}(L-1)\Phi_V^T]$$
$$-\bar{p}_{2,2}(L)[\bar{H}\bar{\Phi}S_{3312}(L-1)\Phi_V^T + \Phi_V^2 S_{3322}(L-1)\Phi_V^T + \Phi_V S_{3332}(L-1)\Phi_V^T]\}\ , \quad S_{2312}(0) = 0 \tag{3.89}$$

$$S_{2313}(L) = \tilde{J}_{21}(L,L)[\ 0 \quad p(L)I_m\ ]R_U(L) + \bar{\Phi}S_{2313}(L-1)\Phi_V^T$$
$$-\tilde{J}_{21}(L,L)\{\bar{p}_{2,2}(L)[\bar{H}\bar{\Phi}S_{1313}(L-1)\Phi_V^T + \Phi_V^2 S_{1323}(L-1)\Phi_V^T + \Phi_V S_{1333}(L-1)\Phi_V^T]$$
$$+\bar{p}(L)[\bar{H}\bar{\Phi}S_{2313}(L-1)\Phi_V^T + \Phi_V^2 S_{2323}(L-1)\Phi_V^T + \Phi_V S_{2333}(L-1)\Phi_V^T]$$
$$-\bar{p}_{2,2}(L)[\bar{H}\bar{\Phi}S_{3313}(L-1)\Phi_V^T + \Phi_V^2 S_{3323}(L-1)\Phi_V^T + \Phi_V S_{3333}(L-1)\Phi_V^T]\}\ , \quad S_{2313}(0) = 0 \tag{3.90}$$

$$S_{2321}(L)=\tilde{J}_{22}(L,L)[\ 0\quad p(L)I_m\ ]\bar{H}K_{\bar{x}}(L,L)+\Phi_V S_{2321}(L-1)\bar{\Phi}^T$$
$$-\tilde{J}_{22}(L,L)\{\bar{p}_{2,2}(L)[\bar{H}\bar{\Phi}S_{1311}(L-1)\bar{\Phi}^T+\Phi_V^2 S_{1321}(L-1)\bar{\Phi}^T+\Phi_V S_{1331}(L-1)\bar{\Phi}^T]$$
$$+\bar{p}(L)[\bar{H}\bar{\Phi}S_{2311}(L-1)\bar{\Phi}^T+\Phi_V^2 S_{2321}(L-1)\bar{\Phi}^T+\Phi_V S_{2331}(L-1)\bar{\Phi}^T]$$
$$-\bar{p}_{2,2}(L)[\bar{H}\bar{\Phi}S_{3311}(L-1)\bar{\Phi}^T+\Phi_V^2 S_{3321}(L-1)\bar{\Phi}^T+\Phi_V S_{3331}(L-1)\bar{\Phi}^T]\},\quad S_{2321}(0)=0$$

(3.91)

$$S_{2322}(L)=\tilde{J}_{22}(L,L)[\ 0\quad p(L)I_m\ ]\Phi_V K_V(L,L)+\Phi_V S_{2322}(L-1)\Phi_V^T$$
$$-\tilde{J}_{22}(L,L)\{\bar{p}_{2,2}(L)[\bar{H}\bar{\Phi}S_{1312}(L-1)\Phi_V^T+\Phi_V^2 S_{1322}(L-1)\Phi_V^T+\Phi_V S_{1332}(L-1)\Phi_V^T]$$
$$+\bar{p}(L)[\bar{H}\bar{\Phi}S_{2312}(L-1)\Phi_V^T+\Phi_V^2 S_{2322}(L-1)\Phi_V^T+\Phi_V S_{2332}(L-1)\Phi_V^T]$$
$$-\bar{p}_{2,2}(L)[\bar{H}\bar{\Phi}S_{3312}(L-1)\Phi_V^T+\Phi_V^2 S_{3322}(L-1)\Phi_V^T+\Phi_V S_{3332}(L-1)\Phi_V^T]\},\quad S_{2322}(0)=0$$

(3.92)

$$S_{2323}(L)=\tilde{J}_{22}(L,L)[\ 0\quad p(L)I_m\ ]R_U(L)+\Phi_V S_{2323}(L-1)\Phi_V^T$$
$$-\tilde{J}_{22}(L,L)\{\bar{p}_{2,2}(L)[\bar{H}\bar{\Phi}S_{1313}(L-1)\Phi_V^T+\Phi_V^2 S_{1323}(L-1)\Phi_V^T+\Phi_V S_{1333}(L-1)\Phi_V^T]$$
$$+\bar{p}(L)[\bar{H}\bar{\Phi}S_{2313}(L-1)\Phi_V^T+\Phi_V^2 S_{2323}(L-1)\Phi_V^T+\Phi_V S_{2333}(L-1)\Phi_V^T]$$
$$-\bar{p}_{2,2}(L)[\bar{H}\bar{\Phi}S_{3313}(L-1)\Phi_V^T+\Phi_V^2 S_{3323}(L-1)\Phi_V^T+\Phi_V S_{3333}(L-1)\Phi_V^T]\},\quad S_{2323}(0)=0$$

(3.93)

$$S_{2331}(L)=\tilde{J}_{23}(L,L)[\ 0\quad p(L)I_m\ ]\bar{H}K_{\bar{x}}(L,L)+\Phi_V S_{2331}(L-1)\bar{\Phi}^T$$
$$-\tilde{J}_{23}(L,L)\{\bar{p}_{2,2}(L)[\bar{H}\bar{\Phi}S_{1311}(L-1)\bar{\Phi}^T+\Phi_V^2 S_{1321}(L-1)\bar{\Phi}^T+\Phi_V S_{1331}(L-1)\bar{\Phi}^T]$$
$$+\bar{p}(L)[\bar{H}\bar{\Phi}S_{2311}(L-1)\bar{\Phi}^T+\Phi_V^2 S_{2321}(L-1)\bar{\Phi}^T+\Phi_V S_{2331}(L-1)\bar{\Phi}^T]$$
$$-\bar{p}_{2,2}(L)[\bar{H}\bar{\Phi}S_{3311}(L-1)\bar{\Phi}^T+\Phi_V^2 S_{3321}(L-1)\bar{\Phi}^T+\Phi_V S_{3331}(L-1)\bar{\Phi}^T]\},\quad S_{2331}(0)=0$$

(3.94)

$$S_{2332}(L)=\tilde{J}_{23}(L,L)[\ 0\quad p(L)I_m\ ]\Phi_V K_V(L,L)+\Phi_V S_{2332}(L-1)\Phi_V^T$$
$$-\tilde{J}_{23}(L,L)\{\bar{p}_{2,2}(L)[\bar{H}\bar{\Phi}S_{1312}(L-1)\Phi_V^T+\Phi_V^2 S_{1322}(L-1)\Phi_V^T+\Phi_V S_{1332}(L-1)\Phi_V^T]$$
$$+\bar{p}(L)[\bar{H}\bar{\Phi}S_{2312}(L-1)\Phi_V^T+\Phi_V^2 S_{2322}(L-1)\Phi_V^T+\Phi_V S_{2332}(L-1)\Phi_V^T]$$
$$-\bar{p}_{2,2}(L)[\bar{H}\bar{\Phi}S_{3312}(L-1)\Phi_V^T+\Phi_V^2 S_{3322}(L-1)\Phi_V^T+\Phi_V S_{3332}(L-1)\Phi_V^T]\},\quad S_{2332}(0)=0$$

(3.95)

$$S_{2333}(L)=\tilde{J}_{23}(L,L)[\ 0\quad p(L)I_m\ ]R_U(L)+\Phi_V S_{2333}(L-1)\Phi_V^T$$
$$-\tilde{J}_{23}(L,L)\{\bar{p}_{2,2}(L)[\bar{H}\bar{\Phi}S_{1313}(L-1)\Phi_V^T+\Phi_V^2 S_{1323}(L-1)\Phi_V^T+\Phi_V S_{1333}(L-1)\Phi_V^T]$$
$$+\bar{p}(L)[\bar{H}\bar{\Phi}S_{2313}(L-1)\Phi_V^T+\Phi_V^2 S_{2323}(L-1)\Phi_V^T+\Phi_V S_{2333}(L-1)\Phi_V^T]$$
$$-\bar{p}_{2,2}(L)[\bar{H}\bar{\Phi}S_{3313}(L-1)\Phi_V^T+\Phi_V^2 S_{3323}(L-1)\Phi_V^T+\Phi_V S_{3333}(L-1)\Phi_V^T]\},\quad S_{2333}(0)=0$$

(3.96)

$$\begin{aligned}S_{3111}(L) &= \tilde{J}_{31}(L,L)[\ p(L)I_m \quad 0\ ]\bar{H}K_{\bar{x}}(L,L) + \bar{\Phi}S_{3111}(L-1)\bar{\Phi}^T \\ &-\tilde{J}_{31}(L,L)\{\bar{p}_{2,2}(L)[\bar{H}\bar{\Phi}S_{1111}(L-1)\bar{\Phi}^T + \Phi_V^2 S_{1121}(L-1)\bar{\Phi}^T + \Phi_V S_{1131}(L-1)\bar{\Phi}^T] \\ &+\bar{p}(L)[\bar{H}\bar{\Phi}S_{2111}(L-1)\bar{\Phi}^T + \Phi_V^2 S_{2121}(L-1)\bar{\Phi}^T + \Phi_V S_{2131}(L-1)\bar{\Phi}^T] \\ &-\bar{p}_{2,2}(L)[\bar{H}\bar{\Phi}S_{3111}(L-1)\bar{\Phi}^T + \Phi_V^2 S_{3121}(L-1)\bar{\Phi}^T + \Phi_V S_{3131}(L-1)\bar{\Phi}^T]\}\ , \quad S_{3111}(0)=0\end{aligned} \tag{3.97}$$

$$\begin{aligned}S_{3112}(L) &= \tilde{J}_{31}(L,L)[\ p(L)I_m \quad 0\ ]\Phi_V K_V(L,L) + \bar{\Phi}S_{3112}(L-1)\Phi_V^T \\ &-\tilde{J}_{31}(L,L)\{\bar{p}_{2,2}(L)[\bar{H}\bar{\Phi}S_{1112}(L-1)\Phi_V^T + \Phi_V^2 S_{1122}(L-1)\Phi_V^T + \Phi_V S_{1132}(L-1)\Phi_V^T] \\ &+\bar{p}(L)[\bar{H}\bar{\Phi}S_{2112}(L-1)\Phi_V^T + \Phi_V^2 S_{2122}(L-1)\Phi_V^T + \Phi_V S_{2132}(L-1)\Phi_V^T] \\ &-\bar{p}_{2,2}(L)[\bar{H}\bar{\Phi}S_{3112}(L-1)\Phi_V^T + \Phi_V^2 S_{3122}(L-1)\Phi_V^T + \Phi_V S_{3132}(L-1)\Phi_V^T]\}\ , \quad S_{3112}(0)=0\end{aligned} \tag{3.98}$$

$$\begin{aligned}S_{3113}(L) &= \tilde{J}_{31}(L,L)[\ p(L)I_m \quad 0\ ]R_U(L) + \bar{\Phi}S_{3113}(L-1)\Phi_V^T \\ &-\tilde{J}_{31}(L,L)\{\bar{p}_{2,2}(L)[\bar{H}\bar{\Phi}S_{1113}(L-1)\Phi_V^T + \Phi_V^2 S_{1123}(L-1)\Phi_V^T + \Phi_V S_{1133}(L-1)\Phi_V^T] \\ &+\bar{p}(L)[\bar{H}\bar{\Phi}S_{2113}(L-1)\Phi_V^T + \Phi_V^2 S_{2123}(L-1)\Phi_V^T + \Phi_V S_{2133}(L-1)\Phi_V^T] \\ &-\bar{p}_{2,2}(L)[\bar{H}\bar{\Phi}S_{3113}(L-1)\Phi_V^T + \Phi_V^2 S_{3123}(L-1)\Phi_V^T + \Phi_V S_{3133}(L-1)\Phi_V^T]\}\ , \quad S_{3113}(0)=0\end{aligned} \tag{3.99}$$

$$\begin{aligned}S_{3121}(L) &= \tilde{J}_{32}(L,L)[\ p(L)I_m \quad 0\ ]\bar{H}K_{\bar{x}}(L,L) + \Phi_V S_{3121}(L-1)\bar{\Phi}^T \\ &-\tilde{J}_{32}(L,L)\{\bar{p}_{2,2}(L)[\bar{H}\bar{\Phi}S_{1111}(L-1)\bar{\Phi}^T + \Phi_V^2 S_{1121}(L-1)\bar{\Phi}^T + \Phi_V S_{1131}(L-1)\bar{\Phi}^T] \\ &+\bar{p}(L)[\bar{H}\bar{\Phi}S_{2111}(L-1)\bar{\Phi}^T + \Phi_V^2 S_{2121}(L-1)\bar{\Phi}^T + \Phi_V S_{2131}(L-1)\bar{\Phi}^T] \\ &-\bar{p}_{2,2}(L)[\bar{H}\bar{\Phi}S_{3111}(L-1)\bar{\Phi}^T + \Phi_V^2 S_{3121}(L-1)\bar{\Phi}^T + \Phi_V S_{3131}(L-1)\bar{\Phi}^T]\}\ , \quad S_{3121}(0)=0\end{aligned} \tag{3.100}$$

$$\begin{aligned}S_{3122}(L) &= \tilde{J}_{32}(L,L)[\ p(L)I_m \quad 0\ ]\Phi_V K_V(L,L) + \Phi_V S_{3122}(L-1)\Phi_V^T \\ &-\tilde{J}_{32}(L,L)\{\bar{p}_{2,2}(L)[\bar{H}\bar{\Phi}S_{1112}(L-1)\Phi_V^T + \Phi_V^2 S_{1122}(L-1)\Phi_V^T + \Phi_V S_{1132}(L-1)\Phi_V^T] \\ &+\bar{p}(L)[\bar{H}\bar{\Phi}S_{2112}(L-1)\Phi_V^T + \Phi_V^2 S_{2122}(L-1)\Phi_V^T + \Phi_V S_{2132}(L-1)\Phi_V^T] \\ &-\bar{p}_{2,2}(L)[\bar{H}\bar{\Phi}S_{3112}(L-1)\Phi_V^T + \Phi_V^2 S_{3122}(L-1)\Phi_V^T + \Phi_V S_{3132}(L-1)\Phi_V^T]\}\ , \quad S_{3122}(0)=0\end{aligned} \tag{3.101}$$

$$\begin{aligned}S_{3123}(L) &= \tilde{J}_{32}(L,L)[\ p(L)I_m \quad 0\ ]R_U(L) + \Phi_V S_{3123}(L-1)\Phi_V^T \\ &-\tilde{J}_{32}(L,L)\{\bar{p}_{2,2}(L)[\bar{H}\bar{\Phi}S_{1113}(L-1)\Phi_V^T + \Phi_V^2 S_{1123}(L-1)\Phi_V^T + \Phi_V S_{1133}(L-1)\Phi_V^T] \\ &+\bar{p}(L)[\bar{H}\bar{\Phi}S_{2113}(L-1)\Phi_V^T + \Phi_V^2 S_{2123}(L-1)\Phi_V^T + \Phi_V S_{2133}(L-1)\Phi_V^T] \\ &-\bar{p}_{2,2}(L)[\bar{H}\bar{\Phi}S_{3113}(L-1)\Phi_V^T + \Phi_V^2 S_{3123}(L-1)\Phi_V^T + \Phi_V S_{3133}(L-1)\Phi_V^T]\}\ , \quad S_{3123}(0)=0\end{aligned} \tag{3.102}$$

$$\begin{aligned}S_{3131}(L) &= \tilde{J}_{33}(L,L)[\ p(L)I_m \quad 0\ ]\bar{H}K_{\bar{x}}(L,L) + \Phi_V S_{3131}(L-1)\bar{\Phi}^T \\ &-\tilde{J}_{33}(L,L)\{\bar{p}_{2,2}(L)[\bar{H}\bar{\Phi}S_{1111}(L-1)\bar{\Phi}^T + \Phi_V^2 S_{1121}(L-1)\bar{\Phi}^T + \Phi_V S_{1131}(L-1)\bar{\Phi}^T] \\ &+\bar{p}(L)[\bar{H}\bar{\Phi}S_{2111}(L-1)\bar{\Phi}^T + \Phi_V^2 S_{2121}(L-1)\bar{\Phi}^T + \Phi_V S_{2131}(L-1)\bar{\Phi}^T] \\ &-\bar{p}_{2,2}(L)[\bar{H}\bar{\Phi}S_{3111}(L-1)\bar{\Phi}^T + \Phi_V^2 S_{3121}(L-1)\bar{\Phi}^T + \Phi_V S_{3131}(L-1)\bar{\Phi}^T]\}\ , \quad S_{3131}(0)=0\end{aligned} \tag{3.103}$$

$$S_{3132}(L) = \tilde{J}_{33}(L,L)[\ p(L)I_m \quad 0\ ]\Phi_V K_V(L,L) + \Phi_V S_{3132}(L-1)\Phi_V^T$$
$$-\tilde{J}_{33}(L,L)\{\bar{p}_{2,2}(L)[\bar{H}\bar{\Phi}S_{1112}(L-1)\Phi_V^T + \Phi_V^2 S_{1122}(L-1)\Phi_V^T + \Phi_V S_{1132}(L-1)\Phi_V^T]$$
$$+\bar{p}(L)[\bar{H}\bar{\Phi}S_{2112}(L-1)\Phi_V^T + \Phi_V^2 S_{2122}(L-1)\Phi_V^T + \Phi_V S_{2132}(L-1)\Phi_V^T]$$
$$-\bar{p}_{2,2}(L)[\bar{H}\bar{\Phi}S_{3112}(L-1)\Phi_V^T + \Phi_V^2 S_{3122}(L-1)\Phi_V^T + \Phi_V S_{3132}(L-1)\Phi_V^T]\}\ , \quad S_{3132}(0) = 0 \tag{3.104}$$

$$S_{3133}(L) = \tilde{J}_{33}(L,L)[\ p(L)I_m \quad 0\ ]R_U(L) + \Phi_V S_{3133}(L-1)\Phi_V^T$$
$$-\tilde{J}_{33}(L,L)\{\bar{p}_{2,2}(L)[\bar{H}\bar{\Phi}S_{1113}(L-1)\Phi_V^T + \Phi_V^2 S_{1123}(L-1)\Phi_V^T + \Phi_V S_{1133}(L-1)\Phi_V^T]$$
$$+\bar{p}(L)[\bar{H}\bar{\Phi}S_{2113}(L-1)\Phi_V^T + \Phi_V^2 S_{2123}(L-1)\Phi_V^T + \Phi_V S_{2133}(L-1)\Phi_V^T]$$
$$-\bar{p}_{2,2}(L)[\bar{H}\bar{\Phi}S_{3113}(L-1)\Phi_V^T + \Phi_V^2 S_{3123}(L-1)\Phi_V^T + \Phi_V S_{3133}(L-1)\Phi_V^T]\}\ , \quad S_{3133}(0) = 0 \tag{3.105}$$

$$S_{3211}(L) = \tilde{J}_{31}(L,L)[\ 0 \quad I_m\ ]\bar{H}K_{\bar{x}}(L,L) + \bar{\Phi}S_{3211}(L-1)\bar{\Phi}^T$$
$$-\tilde{J}_{31}(L,L)\{\bar{p}_{2,2}(L)[\bar{H}\bar{\Phi}S_{1211}(L-1)\bar{\Phi}^T + \Phi_V^2 S_{1221}(L-1)\bar{\Phi}^T + \Phi_V S_{1231}(L-1)\bar{\Phi}^T]$$
$$+\bar{p}(L)[\bar{H}\bar{\Phi}S_{2211}(L-1)\bar{\Phi}^T + \Phi_V^2 S_{2221}(L-1)\bar{\Phi}^T + \Phi_V S_{2231}(L-1)\bar{\Phi}^T]$$
$$-\bar{p}_{2,2}(L)[\bar{H}\bar{\Phi}S_{3211}(L-1)\bar{\Phi}^T + \Phi_V^2 S_{3221}(L-1)\bar{\Phi}^T + \Phi_V S_{3231}(L-1)\bar{\Phi}^T]\}\ , \quad S_{3211}(0) = 0 \tag{3.106}$$

$$S_{3212}(L) = \tilde{J}_{31}(L,L)[\ 0 \quad I_m\ ]\Phi_V K_V(L,L) + \bar{\Phi}S_{2312}(L-1)\Phi_V^T$$
$$-\tilde{J}_{31}(L,L)\{\bar{p}_{2,2}(L)[\bar{H}\bar{\Phi}S_{1212}(L-1)\Phi_V^T + \Phi_V^2 S_{1222}(L-1)\Phi_V^T + \Phi_V S_{1232}(L-1)\Phi_V^T]$$
$$+\bar{p}(L)[\bar{H}\bar{\Phi}S_{2212}(L-1)\Phi_V^T + \Phi_V^2 S_{2222}(L-1)\Phi_V^T + \Phi_V S_{2232}(L-1)\Phi_V^T]$$
$$-\bar{p}_{2,2}(L)[\bar{H}\bar{\Phi}S_{3212}(L-1)\Phi_V^T + \Phi_V^2 S_{3222}(L-1)\Phi_V^T + \Phi_V S_{3232}(L-1)\Phi_V^T]\}\ , \quad S_{3212}(0) = 0 \tag{3.107}$$

$$S_{3213}(L) = \tilde{J}_{31}(L,L)[\ 0 \quad I_m\ ]R_U(L) + \bar{\Phi}S_{3213}(L-1)\Phi_V^T$$
$$-\tilde{J}_{31}(L,L)\{\bar{p}_{2,2}(L)[\bar{H}\bar{\Phi}S_{1213}(L-1)\Phi_V^T + \Phi_V^2 S_{1223}(L-1)\Phi_V^T + \Phi_V S_{1233}(L-1)\Phi_V^T]$$
$$+\bar{p}(L)[\bar{H}\bar{\Phi}S_{2213}(L-1)\Phi_V^T + \Phi_V^2 S_{2223}(L-1)\Phi_V^T + \Phi_V S_{2233}(L-1)\Phi_V^T]$$
$$-\bar{p}_{2,2}(L)[\bar{H}\bar{\Phi}S_{3213}(L-1)\Phi_V^T + \Phi_V^2 S_{3223}(L-1)\Phi_V^T + \Phi_V S_{3233}(L-1)\Phi_V^T]\}\ , \quad S_{3213}(0) = 0 \tag{3.108}$$

$$S_{3221}(L) = \tilde{J}_{32}(L,L)[\ 0 \quad I_m\ ]\bar{H}K_{\bar{x}}(L,L) + \Phi_V S_{3221}(L-1)\bar{\Phi}^T$$
$$-\tilde{J}_{32}(L,L)\{\bar{p}_{2,2}(L)[\bar{H}\bar{\Phi}S_{1211}(L-1)\bar{\Phi}^T + \Phi_V^2 S_{1221}(L-1)\bar{\Phi}^T + \Phi_V S_{1231}(L-1)\bar{\Phi}^T]$$
$$+\bar{p}(L)[\bar{H}\bar{\Phi}S_{2211}(L-1)\bar{\Phi}^T + \Phi_V^2 S_{2221}(L-1)\bar{\Phi}^T + \Phi_V S_{2231}(L-1)\bar{\Phi}^T]$$
$$-\bar{p}_{2,2}(L)[\bar{H}\bar{\Phi}S_{3211}(L-1)\bar{\Phi}^T + \Phi_V^2 S_{3221}(L-1)\bar{\Phi}^T + \Phi_V S_{3231}(L-1)\bar{\Phi}^T]\}\ , \quad S_{3221}(0) = 0 \tag{3.109}$$

$$S_{3222}(L) = \tilde{J}_{32}(L,L)[\ 0 \quad I_m\ ]\Phi_V K_V(L,L) + \Phi_V S_{3222}(L-1)\Phi_V^T$$
$$-\tilde{J}_{32}(L,L)\{\bar{p}_{2,2}(L)[\bar{H}\bar{\Phi}S_{1212}(L-1)\Phi_V^T + \Phi_V^2 S_{1222}(L-1)\Phi_V^T + \Phi_V S_{1232}(L-1)\Phi_V^T]$$
$$+\bar{p}(L)[\bar{H}\bar{\Phi}S_{2212}(L-1)\Phi_V^T + \Phi_V^2 S_{2222}(L-1)\Phi_V^T + \Phi_V S_{2232}(L-1)\Phi_V^T]$$
$$-\bar{p}_{2,2}(L)[\bar{H}\bar{\Phi}S_{3212}(L-1)\Phi_V^T + \Phi_V^2 S_{3222}(L-1)\Phi_V^T + \Phi_V S_{3232}(L-1)\Phi_V^T]\}\ , \quad S_{3222}(0) = 0 \tag{3.110}$$

$$
\begin{aligned}
S_{3223}(L) &= \tilde{J}_{32}(L,L)[\begin{array}{cc} 0 & I_m \end{array}]R_U(L) + \Phi_V S_{3223}(L-1)\Phi_V^T \\
&-\tilde{J}_{32}(L,L)\{\bar{p}_{2,2}(L)[\bar{H}\bar{\Phi}S_{1213}(L-1)\Phi_V^T + \Phi_V^2 S_{1223}(L-1)\Phi_V^T + \Phi_V S_{1233}(L-1)\Phi_V^T] \\
&+\bar{p}(L)[\bar{H}\bar{\Phi}S_{2213}(L-1)\Phi_V^T + \Phi_V^2 S_{2223}(L-1)\Phi_V^T + \Phi_V S_{2233}(L-1)\Phi_V^T] \\
&-\bar{p}_{2,2}(L)[\bar{H}\bar{\Phi}S_{3213}(L-1)\Phi_V^T + \Phi_V^2 S_{3223}(L-1)\Phi_V^T + \Phi_V S_{3233}(L-1)\Phi_V^T]\}, \quad S_{3223}(0)=0
\end{aligned} \tag{3.111}
$$

$$
\begin{aligned}
S_{3231}(L) &= \tilde{J}_{33}(L,L)[\begin{array}{cc} 0 & I_m \end{array}]\bar{H}K_{\tilde{x}}(L,L) + \Phi_V S_{3231}(L-1)\bar{\Phi}^T \\
&-\tilde{J}_{33}(L,L)\{\bar{p}_{2,2}(L)[\bar{H}\bar{\Phi}S_{1211}(L-1)\bar{\Phi}^T + \Phi_V^2 S_{1221}(L-1)\bar{\Phi}^T + \Phi_V S_{1231}(L-1)\bar{\Phi}^T] \\
&+\bar{p}(L)[\bar{H}\bar{\Phi}S_{2211}(L-1)\bar{\Phi}^T + \Phi_V^2 S_{2221}(L-1)\bar{\Phi}^T + \Phi_V S_{2231}(L-1)\bar{\Phi}^T] \\
&-\bar{p}_{2,2}(L)[\bar{H}\bar{\Phi}S_{3211}(L-1)\bar{\Phi}^T + \Phi_V^2 S_{3221}(L-1)\bar{\Phi}^T + \Phi_V S_{3231}(L-1)\bar{\Phi}^T]\}, \quad S_{3231}(0)=0
\end{aligned} \tag{3.112}
$$

$$
\begin{aligned}
S_{3232}(L) &= \tilde{J}_{33}(L,L)[\begin{array}{cc} 0 & I_m \end{array}]\Phi_V K_V(L,L) + \Phi_V S_{3232}(L-1)\Phi_V^T \\
&-\tilde{J}_{33}(L,L)\{\bar{p}_{2,2}(L)[\bar{H}\bar{\Phi}S_{1212}(L-1)\Phi_V^T + \Phi_V^2 S_{1222}(L-1)\Phi_V^T + \Phi_V S_{1232}(L-1)\Phi_V^T] \\
&+\bar{p}(L)[\bar{H}\bar{\Phi}S_{2212}(L-1)\Phi_V^T + \Phi_V^2 S_{2222}(L-1)\Phi_V^T + \Phi_V S_{2232}(L-1)\Phi_V^T] \\
&-\bar{p}_{2,2}(L)[\bar{H}\bar{\Phi}S_{3212}(L-1)\Phi_V^T + \Phi_V^2 S_{3222}(L-1)\Phi_V^T + \Phi_V S_{3232}(L-1)\Phi_V^T]\}, \quad S_{3232}(0)=0
\end{aligned} \tag{3.113}
$$

$$
\begin{aligned}
S_{3233}(L) &= \tilde{J}_{33}(L,L)[\begin{array}{cc} 0 & I_m \end{array}]R_U(L) + \Phi_V S_{3233}(L-1)\Phi_V^T \\
&-\tilde{J}_{33}(L,L)\{\bar{p}_{2,2}(L)[\bar{H}\bar{\Phi}S_{1213}(L-1)\Phi_V^T + \Phi_V^2 S_{1223}(L-1)\Phi_V^T + \Phi_V S_{1233}(L-1)\Phi_V^T] \\
&+\bar{p}(L)[\bar{H}\bar{\Phi}S_{2213}(L-1)\Phi_V^T + \Phi_V^2 S_{2223}(L-1)\Phi_V^T + \Phi_V S_{2233}(L-1)\Phi_V^T] \\
&-\bar{p}_{2,2}(L)[\bar{H}\bar{\Phi}S_{3213}(L-1)\Phi_V^T + \Phi_V^2 S_{3223}(L-1)\Phi_V^T + \Phi_V S_{3233}(L-1)\Phi_V^T]\}, \quad S_{3233}(0)=0
\end{aligned} \tag{3.114}
$$

$$
\begin{aligned}
S_{3311}(L) &= \tilde{J}_{31}(L,L)[\begin{array}{cc} 0 & p(L)I_m \end{array}]\bar{H}K_{\tilde{x}}(L,L) + \bar{\Phi}S_{3311}(L-1)\bar{\Phi}^T \\
&-\tilde{J}_{31}(L,L)\{\bar{p}_{2,2}(L)[\bar{H}\bar{\Phi}S_{1311}(L-1)\bar{\Phi}^T + \Phi_V^2 S_{1321}(L-1)\bar{\Phi}^T + \Phi_V S_{1331}(L-1)\bar{\Phi}^T] \\
&+\bar{p}(L)[\bar{H}\bar{\Phi}S_{2311}(L-1)\bar{\Phi}^T + \Phi_V^2 S_{2321}(L-1)\bar{\Phi}^T + \Phi_V S_{2331}(L-1)\bar{\Phi}^T] \\
&-\bar{p}_{2,2}(L)[\bar{H}\bar{\Phi}S_{3311}(L-1)\bar{\Phi}^T + \Phi_V^2 S_{3321}(L-1)\bar{\Phi}^T + \Phi_V S_{3331}(L-1)\bar{\Phi}^T]\}, \quad S_{3311}(0)=0
\end{aligned} \tag{3.115}
$$

$$
\begin{aligned}
S_{3312}(L) &= \tilde{J}_{31}(L,L)[\begin{array}{cc} 0 & p(L)I_m \end{array}]\Phi_V K_V(L,L) + \bar{\Phi}S_{3312}(L-1)\Phi_V^T \\
&-\tilde{J}_{31}(L,L)\{\bar{p}_{2,2}(L)[\bar{H}\bar{\Phi}S_{1312}(L-1)\Phi_V^T + \Phi_V^2 S_{1322}(L-1)\Phi_V^T + \Phi_V S_{1332}(L-1)\Phi_V^T] \\
&+\bar{p}(L)[\bar{H}\bar{\Phi}S_{2312}(L-1)\Phi_V^T + \Phi_V^2 S_{2322}(L-1)\Phi_V^T + \Phi_V S_{2332}(L-1)\Phi_V^T] \\
&-\bar{p}_{2,2}(L)[\bar{H}\bar{\Phi}S_{3312}(L-1)\Phi_V^T + \Phi_V^2 S_{3322}(L-1)\Phi_V^T + \Phi_V S_{3332}(L-1)\Phi_V^T]\}, \quad S_{3312}(0)=0
\end{aligned} \tag{3.116}
$$

$$
\begin{aligned}
S_{3313}(L) &= \tilde{J}_{31}(L,L)[\begin{array}{cc} 0 & p(L)I_m \end{array}]R_U(L) + \bar{\Phi}S_{3313}(L-1)\Phi_V^T \\
&-\tilde{J}_{31}(L,L)\{\bar{p}_{2,2}(L)[\bar{H}\bar{\Phi}S_{1313}(L-1)\Phi_V^T + \Phi_V^2 S_{1323}(L-1)\Phi_V^T + \Phi_V S_{1333}(L-1)\Phi_V^T] \\
&+\bar{p}(L)[\bar{H}\bar{\Phi}S_{2313}(L-1)\Phi_V^T + \Phi_V^2 S_{2323}(L-1)\Phi_V^T + \Phi_V S_{2333}(L-1)\Phi_V^T] \\
&-\bar{p}_{2,2}(L)[\bar{H}\bar{\Phi}S_{3313}(L-1)\Phi_V^T + \Phi_V^2 S_{3323}(L-1)\Phi_V^T + \Phi_V S_{3333}(L-1)\Phi_V^T]\}, \quad S_{3313}(0)=0
\end{aligned} \tag{3.117}
$$

$$S_{3321}(L) = \tilde{J}_{32}(L,L)[\ 0 \quad p(L)I_m\ ]\bar{H}K_{\bar{x}}(L,L) + \Phi_V S_{3321}(L-1)\bar{\Phi}^T$$
$$-\tilde{J}_{32}(L,L)\{\bar{p}_{2,2}(L)[\bar{H}\bar{\Phi}S_{1311}(L-1)\bar{\Phi}^T + \Phi_V^2 S_{1321}(L-1)\bar{\Phi}^T + \Phi_V S_{1331}(L-1)\bar{\Phi}^T]$$
$$+\bar{p}(L)[\bar{H}\bar{\Phi}S_{2311}(L-1)\bar{\Phi}^T + \Phi_V^2 S_{2321}(L-1)\bar{\Phi}^T + \Phi_V S_{2331}(L-1)\bar{\Phi}^T]$$
$$-\bar{p}_{2,2}(L)[\bar{H}\bar{\Phi}S_{3311}(L-1)\bar{\Phi}^T + \Phi_V^2 S_{3321}(L-1)\bar{\Phi}^T + \Phi_V S_{3331}(L-1)\bar{\Phi}^T]\}\ , \quad S_{3321}(0) = 0 \tag{3.118}$$

$$S_{3322}(L) = \tilde{J}_{32}(L,L)[\ 0 \quad p(L)I_m\ ]\Phi_V K_V(L,L) + \Phi_V S_{3322}(L-1)\Phi_V^T$$
$$-\tilde{J}_{32}(L,L)\{\bar{p}_{2,2}(L)[\bar{H}\bar{\Phi}S_{1312}(L-1)\Phi_V^T + \Phi_V^2 S_{1322}(L-1)\Phi_V^T + \Phi_V S_{1332}(L-1)\Phi_V^T]$$
$$+\bar{p}(L)[\bar{H}\bar{\Phi}S_{2312}(L-1)\Phi_V^T + \Phi_V^2 S_{2322}(L-1)\Phi_V^T + \Phi_V S_{2332}(L-1)\Phi_V^T]$$
$$-\bar{p}_{2,2}(L)[\bar{H}\bar{\Phi}S_{3312}(L-1)\Phi_V^T + \Phi_V^2 S_{3322}(L-1)\Phi_V^T + \Phi_V S_{3332}(L-1)\Phi_V^T]\}\ , \quad S_{3322}(0) = 0 \tag{3.119}$$

$$S_{3323}(L) = \tilde{J}_{32}(L,L)[\ 0 \quad p(L)I_m\ ]R_U(L) + \Phi_V S_{3323}(L-1)\Phi_V^T$$
$$-\tilde{J}_{32}(L,L)\{\bar{p}_{2,2}(L)[\bar{H}\bar{\Phi}S_{1313}(L-1)\Phi_V^T + \Phi_V^2 S_{1323}(L-1)\Phi_V^T + \Phi_V S_{1333}(L-1)\Phi_V^T]$$
$$+\bar{p}(L)[\bar{H}\bar{\Phi}S_{2313}(L-1)\Phi_V^T + \Phi_V^2 S_{2323}(L-1)\Phi_V^T + \Phi_V S_{2333}(L-1)\Phi_V^T]$$
$$-\bar{p}_{2,2}(L)[\bar{H}\bar{\Phi}S_{3313}(L-1)\Phi_V^T + \Phi_V^2 S_{3323}(L-1)\Phi_V^T + \Phi_V S_{3333}(L-1)\Phi_V^T]\}\ , \quad S_{3323}(0) = 0 \tag{3.120}$$

$$S_{3331}(L) = \tilde{J}_{33}(L,L)[\ 0 \quad p(L)I_m\ ]\bar{H}K_{\bar{x}}(L,L) + \Phi_V S_{3331}(L-1)\bar{\Phi}^T$$
$$-\tilde{J}_{33}(L,L)\{\bar{p}_{2,2}(L)[\bar{H}\bar{\Phi}S_{1311}(L-1)\bar{\Phi}^T + \Phi_V^2 S_{1321}(L-1)\bar{\Phi}^T + \Phi_V S_{1331}(L-1)\bar{\Phi}^T]$$
$$+\bar{p}(L)[\bar{H}\bar{\Phi}S_{2311}(L-1)\bar{\Phi}^T + \Phi_V^2 S_{2321}(L-1)\bar{\Phi}^T + \Phi_V S_{2331}(L-1)\bar{\Phi}^T]$$
$$-\bar{p}_{2,2}(L)[\bar{H}\bar{\Phi}S_{3311}(L-1)\bar{\Phi}^T + \Phi_V^2 S_{3321}(L-1)\bar{\Phi}^T + \Phi_V S_{3331}(L-1)\bar{\Phi}^T]\}\ , \quad S_{3331}(0) = 0 \tag{3.121}$$

$$S_{3332}(L) = \tilde{J}_{33}(L,L)[\ 0 \quad p(L)I_m\ ]\Phi_V K_V(L,L) + \Phi_V S_{3332}(L-1)\Phi_V^T$$
$$-\tilde{J}_{33}(L,L)\{\bar{p}_{2,2}(L)[\bar{H}\bar{\Phi}S_{1312}(L-1)\Phi_V^T + \Phi_V^2 S_{1322}(L-1)\Phi_V^T + \Phi_V S_{1332}(L-1)\Phi_V^T]$$
$$+\bar{p}(L)[\bar{H}\bar{\Phi}S_{2312}(L-1)\Phi_V^T + \Phi_V^2 S_{2322}(L-1)\Phi_V^T + \Phi_V S_{2332}(L-1)\Phi_V^T]$$
$$-\bar{p}_{2,2}(L)[\bar{H}\bar{\Phi}S_{3312}(L-1)\Phi_V^T + \Phi_V^2 S_{3322}(L-1)\Phi_V^T + \Phi_V S_{3332}(L-1)\Phi_V^T]\}\ , \quad S_{3332}(0) = 0 \tag{3.122}$$

$$S_{3333}(L) = \tilde{J}_{33}(L,L)[\ 0 \quad p(L)I_m\ ]R_U(L) + \Phi_V S_{3333}(L-1)\Phi_V^T$$
$$-\tilde{J}_{33}(L,L)\{\bar{p}_{2,2}(L)[\bar{H}\bar{\Phi}S_{1313}(L-1)\Phi_V^T + \Phi_V^2 S_{1323}(L-1)\Phi_V^T + \Phi_V S_{1333}(L-1)\Phi_V^T]$$
$$+\bar{p}(L)[\bar{H}\bar{\Phi}S_{2313}(L-1)\Phi_V^T + \Phi_V^2 S_{2323}(L-1)\Phi_V^T + \Phi_V S_{2333}(L-1)\Phi_V^T]$$
$$-\bar{p}_{2,2}(L)[\bar{H}\bar{\Phi}S_{3313}(L-1)\Phi_V^T + \Phi_V^2 S_{3323}(L-1)\Phi_V^T + \Phi_V S_{3333}(L-1)\Phi_V^T]\}\ , \quad S_{3333}(0) = 0 \tag{3.123}$$

Filtering estimate of the state vector: $\hat{\hat{x}}(L,L)$

$$\hat{\hat{x}}(L,L) = \bar{\Phi}\hat{\hat{x}}(L-1,L-1) + \bar{G}(L,L)\{y(L) - \bar{p}_{2,2}(L)[\bar{H}\bar{\Phi}\bar{O}_{11}(L-1)$$
$$+\Phi_V^2\bar{O}_{12}(L-1) + \Phi_V\bar{O}_{13}(L-1)]$$
$$-\bar{p}(L)[\bar{H}\bar{\Phi}\bar{O}_{21}(L-1) + \Phi_V^2\bar{O}_{22}(L-1) + \Phi_V\bar{O}_{23}(L-1]$$
$$+\bar{p}_{2,2}(L)[\bar{H}\bar{\Phi}\bar{O}_{31}(L-1) + \Phi_V^2\bar{O}_{32}(L-1) + \Phi_V\bar{O}_{33}(L-1]\} \ , \quad \hat{\hat{x}}(0,0) = 0 \tag{3.124}$$

Filter gain for $\hat{x}(L,L)$**:** $\bar{G}(L,L)$

$$\bar{G}(L,L) = \{K_{\bar{x}}(L,L)\bar{H}^T\bar{p}^T(L) - [\bar{\Phi}\tilde{B}_{11}(L-1)\bar{\Phi}^T\bar{H}^T$$
$$+\bar{\Phi}\tilde{B}_{12}(L-1)(\Phi_V^T)^2 + \bar{\Phi}\tilde{B}_{13}(L-1)\Phi_V^T]\bar{p}_{2,2}^T(L)$$
$$-[\bar{\Phi}\tilde{B}_{21}(L-1)\bar{\Phi}^T\bar{H}^T + \bar{\Phi}\tilde{B}_{22}(L-1)(\Phi_V^T)^2$$
$$+\bar{\Phi}\tilde{B}_{23}(L-1)\Phi_V^T]\bar{p}^T(L) + [\bar{\Phi}\tilde{B}_{31}(L-1)\bar{\Phi}^T\bar{H}^T + \bar{\Phi}\tilde{B}_{32}(L-1)(\Phi_V^T)^2$$
$$+\bar{\Phi}\tilde{B}_{33}(L-1)\Phi_V^T]\bar{p}_{2,2}^T(L)\}\Gamma^{-1}(L) \tag{3.125}$$

$$\tilde{B}_{11}(L) = \bar{G}(L,L)[\ p(L)I_m \quad 0\ ]\bar{H}K_{\bar{x}}(L,L) + \bar{\Phi}\tilde{B}_{11}(L-1)\bar{\Phi}^T$$
$$-\bar{G}(L,L)\{\bar{p}_{2,2}(L)[\bar{H}\bar{\Phi}S_{1111}(L-1)\bar{\Phi}^T + \Phi_V^2 S_{1121}(L-1)\bar{\Phi}^T + \Phi_V S_{1131}(L-1)\bar{\Phi}^T]$$
$$+\bar{p}(L)[\bar{H}\bar{\Phi}S_{2111}(L-1)\bar{\Phi}^T + \Phi_V^2 S_{2121}(L-1)\bar{\Phi}^T + \Phi_V S_{2131}(L-1)\bar{\Phi}^T]$$
$$-\bar{p}_{2,2}(L)[\bar{H}\bar{\Phi}S_{3111}(L-1)\bar{\Phi}^T + \Phi_V^2 S_{3121}(L-1)\bar{\Phi}^T + \Phi_V S_{3131}(L-1)\bar{\Phi}^T]\} \ , \quad \tilde{B}_{11}(0) = 0 \tag{3.126}$$

$$\tilde{B}_{12}(L) = \bar{G}(L,L)[\ p(L)I_m \quad 0\ ]\Phi_V K_V(L,L) + \bar{\Phi}\tilde{B}_{12}(L-1)\Phi_V^T$$
$$-\bar{G}(L,L)\{\bar{p}_{2,2}(L)[\bar{H}\bar{\Phi}S_{1112}(L-1)\Phi_V^T + \Phi_V^2 S_{1122}(L-1)\Phi_V^T + \Phi_V S_{1132}(L-1)\Phi_V^T]$$
$$+\bar{p}(L)[\bar{H}\bar{\Phi}S_{2112}(L-1)\Phi_V^T + \Phi_V^2 S_{2122}(L-1)\Phi_V^T + \Phi_V S_{2132}(L-1)\Phi_V^T]$$
$$-\bar{p}_{2,2}(L)[\bar{H}\bar{\Phi}S_{3112}(L-1)\Phi_V^T + \Phi_V^2 S_{3122}(L-1)\Phi_V^T + \Phi_V S_{3132}(L-1)\Phi_V^T]\} \ , \quad \tilde{B}_{12}(0) = 0 \tag{3.127}$$

$$\tilde{B}_{13}(L) = \bar{G}(L,L)[\ p(L)I_m \quad 0\ ]R_U(L) + \bar{\Phi}\tilde{B}_{13}(L-1)\Phi_V^T$$
$$-\bar{G}(L,L)\{\bar{p}_{2,2}(L)[\bar{H}\bar{\Phi}S_{1113}(L-1)\Phi_V^T + \Phi_V^2 S_{1123}(L-1)\Phi_V^T + \Phi_V S_{1133}(L-1)\Phi_V^T]$$
$$+\bar{p}(L)[\bar{H}\bar{\Phi}S_{2113}(L-1)\Phi_V^T + \Phi_V^2 S_{2123}(L-1)\Phi_V^T + \Phi_V S_{2133}(L-1)\Phi_V^T]$$
$$-\bar{p}_{2,2}(L)[\bar{H}\bar{\Phi}S_{3113}(L-1)\Phi_V^T + \Phi_V^2 S_{3123}(L-1)\Phi_V^T + \Phi_V S_{3133}(L-1)\Phi_V^T]\} \ , \quad \tilde{B}_{13}(0) = 0 \tag{3.128}$$

$$\tilde{B}_{21}(L) = \bar{G}(L,L)[\ 0 \quad I_m\ ]\bar{H}K_{\bar{x}}(L,L) + \bar{\Phi}\tilde{B}_{21}(L-1)\bar{\Phi}^T$$
$$-\bar{G}(L,L)\{\bar{p}_{2,2}(L)[\bar{H}\bar{\Phi}S_{1211}(L-1)\bar{\Phi}^T + \Phi_V^2 S_{1221}(L-1)\bar{\Phi}^T + \Phi_V S_{1231}(L-1)\bar{\Phi}^T]$$
$$+\bar{p}(L)[\bar{H}\bar{\Phi}S_{2211}(L-1)\bar{\Phi}^T + \Phi_V^2 S_{2221}(L-1)\bar{\Phi}^T + \Phi_V S_{2231}(L-1)\bar{\Phi}^T]$$
$$-\bar{p}_{2,2}(L)[\bar{H}\bar{\Phi}S_{3211}(L-1)\bar{\Phi}^T + \Phi_V^2 S_{3221}(L-1)\bar{\Phi}^T + \Phi_V S_{3231}(L-1)\bar{\Phi}^T]\} \ , \quad \tilde{B}_{21}(0) = 0 \tag{3.129}$$

$$\tilde{B}_{22}(L) = \bar{G}(L,L)[\ 0 \quad I_m\ ]\Phi_V K_V(L,L) + \bar{\Phi}\tilde{B}_{22}(L-1)\Phi_V^T$$
$$-\bar{G}(L,L)\{\bar{p}_{2,2}(L)[\bar{H}\bar{\Phi}S_{1212}(L-1)\Phi_V^T + \Phi_V^2 S_{1222}(L-1)\Phi_V^T + \Phi_V S_{1232}(L-1)\Phi_V^T]$$
$$+\bar{p}(L)[\bar{H}\bar{\Phi}S_{2212}(L-1)\Phi_V^T + \Phi_V^2 S_{2222}(L-1)\Phi_V^T + \Phi_V S_{2232}(L-1)\Phi_V^T]$$
$$-\bar{p}_{2,2}(L)[\bar{H}\bar{\Phi}S_{3212}(L-1)\Phi_V^T + \Phi_V^2 S_{3222}(L-1)\Phi_V^T + \Phi_V S_{3232}(L-1)\Phi_V^T]\}\ , \quad \tilde{B}_{22}(0) = 0 \tag{3.130}$$

$$\tilde{B}_{23}(L) = \bar{G}(L,L)[\ 0 \quad I_m\ ]R_U(L) + \bar{\Phi}\tilde{B}_{23}(L-1)\Phi_V^T$$
$$-\bar{G}(L,L)\{\bar{p}_{2,2}(L)[\bar{H}\bar{\Phi}S_{1213}(L-1)\Phi_V^T + \Phi_V^2 S_{1223}(L-1)\Phi_V^T + \Phi_V S_{1233}(L-1)\Phi_V^T]$$
$$+\bar{p}(L)[\bar{H}\bar{\Phi}S_{2213}(L-1)\Phi_V^T + \Phi_V^2 S_{2223}(L-1)\Phi_V^T + \Phi_V S_{2233}(L-1)\Phi_V^T]$$
$$-\bar{p}_{2,2}(L)[\bar{H}\bar{\Phi}S_{3213}(L-1)\Phi_V^T + \Phi_V^2 S_{3223}(L-1)\Phi_V^T + \Phi_V S_{3233}(L-1)\Phi_V^T]\}\ , \quad \tilde{B}_{23}(0) = 0 \tag{3.131}$$

$$\tilde{B}_{31}(L) = \bar{G}(L,L)[\ 0 \quad p(L)I_m\ ]\bar{H}K_{\bar{x}}(L,L) + \bar{\Phi}\tilde{B}_{31}(L-1)\bar{\Phi}^T$$
$$-\bar{G}(L,L)\{\bar{p}_{2,2}(L)[\bar{H}\bar{\Phi}S_{1311}(L-1)\bar{\Phi}^T + \Phi_V^2 S_{1321}(L-1)\bar{\Phi}^T + \Phi_V S_{1331}(L-1)\bar{\Phi}^T]$$
$$+\bar{p}(L)[\bar{H}\bar{\Phi}S_{2311}(L-1)\bar{\Phi}^T + \Phi_V^2 S_{2321}(L-1)\bar{\Phi}^T + \Phi_V S_{2331}(L-1)\bar{\Phi}^T]$$
$$-\bar{p}_{2,2}(L)[\bar{H}\bar{\Phi}S_{3311}(L-1)\bar{\Phi}^T + \Phi_V^2 S_{3321}(L-1)\bar{\Phi}^T + \Phi_V S_{3331}(L-1)\bar{\Phi}^T]\}\ , \quad \tilde{B}_{31}(0) = 0 \tag{3.132}$$

$$\tilde{B}_{32}(L) = \bar{G}(L,L)[\ 0 \quad p(L)I_m\ ]\Phi_V K_V(L,L) + \bar{\Phi}\tilde{B}_{32}(L-1)\Phi_V^T$$
$$-\bar{G}(L,L)\{\bar{p}_{2,2}(L)[\bar{H}\bar{\Phi}S_{1312}(L-1)\Phi_V^T + \Phi_V^2 S_{1322}(L-1)\Phi_V^T + \Phi_V S_{1332}(L-1)\Phi_V^T]$$
$$+\bar{p}(L)[\bar{H}\bar{\Phi}S_{2312}(L-1)\Phi_V^T + \Phi_V^2 S_{2322}(L-1)\Phi_V^T + \Phi_V S_{2332}(L-1)\Phi_V^T]$$
$$-\bar{p}_{2,2}(L)[\bar{H}\bar{\Phi}S_{3312}(L-1)\Phi_V^T + \Phi_V^2 S_{3322}(L-1)\Phi_V^T + \Phi_V S_{3332}(L-1)\Phi_V^T]\}\ , \quad \tilde{B}_{32}(0) = 0 \tag{3.133}$$

$$\tilde{B}_{33}(L) = \bar{G}(L,L)[\ 0 \quad p(L)I_m\ ]R_U(L) + \bar{\Phi}\tilde{B}_{33}(L-1)\Phi_V^T$$
$$-\bar{G}(L,L)\{\bar{p}_{2,2}(L)[\bar{H}\bar{\Phi}S_{1313}(L-1)\Phi_V^T + \Phi_V^2 S_{1323}(L-1)\Phi_V^T + \Phi_V S_{1333}(L-1)\Phi_V^T]$$
$$+\bar{p}(L)[\bar{H}\bar{\Phi}S_{2313}(L-1)\Phi_V^T + \Phi_V^2 S_{2323}(L-1)\Phi_V^T + \Phi_V S_{2333}(L-1)\Phi_V^T]$$
$$-\bar{p}_{2,2}(L)[\bar{H}\bar{\Phi}S_{3313}(L-1)\Phi_V^T + \Phi_V^2 S_{3323}(L-1)\Phi_V^T + \Phi_V S_{3333}(L-1)\Phi_V^T]\}\ , \quad \tilde{B}_{33}(0) = 0 \tag{3.134}$$

Proof of Theorem 1 is deferred to the Appendix.

From (3.124), it is seen that the innovation process is represented by

$$\upsilon(L) = y(L) - \hat{y}(L,L-1),$$

$$\hat{y}(L,L-1) = \bar{p}_{2,2}(L)[\bar{H}\bar{\Phi}\bar{O}_{11}(L-1) + \Phi_V^2\bar{O}_{12}(L-1) + \Phi_V\bar{O}_{13}(L-1)]$$
$$+\bar{p}(L)[\bar{H}\bar{\Phi}\bar{O}_{21}(L-1) + \Phi_V^2\bar{O}_{22}(L-1) + \Phi_V\bar{O}_{23}(L-1]$$
$$-\bar{p}_{2,2}(L)[\bar{H}\bar{\Phi}\bar{O}_{31}(L-1) + \Phi_V^2\bar{O}_{32}(L-1) + \Phi_V\bar{O}_{33}(L-1]. \tag{3.135}$$

Chapter 4

A Numerical Simulation Example

Let an m-dimensional observation equation be given by

$$\begin{aligned} y(k) &= (1-\gamma(k))\bar{y}(k) + \gamma(k)\bar{y}(k-1), \quad E_\gamma[\gamma(k)] = p(k), \\ \bar{y}(k) &= z(k) + v(k), \quad z(k) = Hx(k), \quad E[v(k)v^T(s)] = R\delta_K(k-s). \end{aligned} \tag{4.1}$$

Let the observation noise $v(k)$ be zero-mean white Gaussian process with the variance R, $N(0,R)$. (4.1) is also written as

$$\begin{aligned} y(k) &= \bar{\gamma}(k)\bar{z}(k) + \bar{v}(k), \\ \bar{\gamma}(k) &= \begin{bmatrix} \gamma(k) & (1-\gamma(k)) \end{bmatrix}, \quad \bar{v}(k) = \bar{\gamma}(k)\begin{bmatrix} v(k-1) & v(k) \end{bmatrix}^T, \\ \bar{z}(k) &= \begin{bmatrix} z(k-1) \\ z(k) \end{bmatrix} = \begin{bmatrix} Hx(k-1) \\ Hx(k) \end{bmatrix} = \bar{H}\bar{x}(k). \end{aligned} \tag{4.2}$$

As in [14], we consider that the signal $z(k)$ is transmitted through one of two channels, each characterized by its observation equation as follows:
Channel 1: $y(k) = \bar{y}(k-1) + v(k)$,
Channel 2: $y(k) = (1 - U(k)\bar{y}(k) + v(k)$,

where $\{v(k)\}$ is a zero-mean white observation noise and $\{U(k)\}$ is a sequence of independent random variables taking values 0 or 1 with
$P\{U(k) = 1\} = \breve{p} = 0.03$, for all k.

We assume that channel 1 is chosen at random with probability $1-q=0.02$ and, hence, channel 2 is selected with probability $q=0.98$. Then, the observation equation is described by $y(k)=\gamma(k)\bar{z}(k)+\bar{v}(k)$, where $\gamma(k)=(1-\alpha)1+\alpha U(k)$ and α is a random variable, independent of $\{U(k)\}$, taking values 0 or 1 with $P\{\alpha=1\}=q=0.98$. It is clear that $\{\gamma(k)\}$ is a sequence of random variables which take values 0 or 1 with

$$\begin{aligned} p(k) &= P\{\gamma(k)=1\}=P\{\alpha=1,U(k)=1\}+P\{\alpha=0\} \\ &= \breve{p}q+(1-q)=0.0397 \end{aligned}$$

for all k, and conditional probability matrix

$$\begin{aligned} P(k|j) &= \begin{bmatrix} 1-\breve{p} & \breve{p} \\ \frac{q\breve{p}(1-\breve{p})}{1-q(1-\breve{p})} & \frac{1-q(1-\breve{p}^2)}{1-q(1-\breve{p})} \end{bmatrix} \\ &= \begin{bmatrix} 0.97 & 0.03 \\ 0.7257 & 0.2743 \end{bmatrix}, \end{aligned}$$

for all $k, j=0,\ldots,k-1$.From (3), $[P(k|j)]_{2,2}=P_{2,2}(k)=0.2743$, for all $k, j=0,\ldots,k-1$.

Let the signal $z(k)$ be generated by the second-order AR model.

$$\begin{aligned} &z(k+1)=-a_1z(k)-a_2z(k-1)+w(k), \quad E[w(k)w(s)]=\sigma^2\delta_K(k-s), \\ &a_1=-0.1, \quad a_2=-0.8, \quad \sigma=0.5. \end{aligned} \tag{4.3}$$

The state-space model for $z(k)$ is given by

$$\begin{aligned} &z(k)=Hx(k)=x_1(k), \quad H=\begin{bmatrix} 1 & 0 \end{bmatrix}, \quad x(k)=\begin{bmatrix} x_1(k) \\ x_2(k) \end{bmatrix}, \\ &\begin{bmatrix} x_1(k+1) \\ x_2(k+1) \end{bmatrix}=\begin{bmatrix} 0 & 1 \\ -a_2 & -a_1 \end{bmatrix}\begin{bmatrix} x_1(k) \\ x_2(k) \end{bmatrix}+\begin{bmatrix} 0 \\ 1 \end{bmatrix}w(k). \end{aligned} \tag{4.4}$$

Hence, $\bar{H}$ and $\bar{x}(k)$ are given by

$$\bar{H}=\begin{bmatrix} H & 0 \\ 0 & H \end{bmatrix}=\begin{bmatrix} 1 & 0 & 0 & 0 \\ 0 & 0 & 1 & 0 \end{bmatrix}, \quad \bar{x}(k)=\begin{bmatrix} x_1(k-1) \\ x_2(k-1) \\ x_1(k) \\ x_2(k) \end{bmatrix}=\begin{bmatrix} x_1(k-1) \\ x_1(k) \\ x_1(k) \\ x_1(k+1) \end{bmatrix}. \tag{4.5}$$

From $\bar{x}(k)$, the signal $z(k)$ is calculated by

$$z(k) = \begin{bmatrix} 0 & 0 & 1 & 0 \end{bmatrix} \bar{x}(k) = \begin{bmatrix} 0 & 1 & 0 & 0 \end{bmatrix} \bar{x}(k) = x_1(k). \tag{4.6}$$

The auto-covariance function of the signal $z(k)$ is given by [14]. From

$$\begin{aligned} &K(0) = \sigma^2, \\ &K(m) = \sigma^2\{\alpha_1(\alpha_2^2-1)\alpha_1^m/[(\alpha_2-\alpha_1)(\alpha_2\alpha_1+1)] \\ &-\alpha_2(\alpha_1^2-1)\alpha_2^m/[(\alpha_2-\alpha_1)(\alpha_1\alpha_2+1)]\}, \quad 0 < m, \\ &\alpha_1, \alpha_2 = (-a_1 \pm \sqrt{a_1^2-4a_2})/2, \end{aligned} \tag{4.7}$$

$$K_x(k,k) = \begin{bmatrix} K(0) & K(1) \\ K(1) & K(0) \end{bmatrix}, \quad \Phi = \begin{bmatrix} 0 & 1 \\ -a_2 & -a_1 \end{bmatrix},$$

$\bar{\Phi}$ and $K_{\bar{x}}(0)$ are given by

$$\bar{\Phi} = \begin{bmatrix} 0 & I \\ 0 & \Phi \end{bmatrix} = \begin{bmatrix} 0 & 0 & 1 & 0 \\ 0 & 0 & 0 & 1 \\ 0 & 0 & 0 & 1 \\ 0 & 0 & -a_2 & -a_1 \end{bmatrix}, \tag{4.8}$$

$$\begin{aligned} K_{\bar{x}}(0) &= \begin{bmatrix} K_x(0) & K_x(0)\Phi^T \\ \Phi K_x(0) & K_x(0) \end{bmatrix} \\ &= \begin{bmatrix} K(0) & K(1) & K(1) & -a_2\cdot K(0)-a_1\cdot K(1) \\ K(1) & K(0) & K(0) & -a_2\cdot K(1)-a_1\cdot K(0) \\ K(1) & K(0) & K(0) & K(1) \\ -a_2\cdot K(0)-a_1\cdot K(1) & -a_2\cdot K(1)-a_1\cdot K(0) & K(1) & K(0) \end{bmatrix}, \end{aligned}$$

$$K(0) = 0.25, \quad K(1) = 0.125. \tag{4.9}$$

$$\Phi_V = \begin{bmatrix} 0 & I \\ 0 & 0 \end{bmatrix}, \quad K_V(k,k) = \begin{bmatrix} R & 0 \\ 0 & R \end{bmatrix} \tag{4.10}$$

From (2.24), (4.2), $\bar{H}$ in (4.5), $\bar{p}(L) = E_\gamma\begin{bmatrix} \gamma(L) & 1-\gamma(L) \end{bmatrix} = \begin{bmatrix} p(L) & 1-p(L) \end{bmatrix}$, $\bar{R}(L)$ is calculated as

$$\begin{aligned} \bar{R}(L) &= E_\gamma[\bar{\gamma}(L)R_U(L)\bar{\gamma}^T(L)] \\ &= E_\gamma\{[\ \gamma(L) \quad (1-\gamma(L)\ ](K_V(k,k)-\Phi_V K_V(k,k)\Phi_V^T)[\ \gamma(L) \quad (1-\gamma(L)\ ]^T\} \\ &= E_\gamma\{[\ \gamma(L) \quad (1-\gamma(L)\ ]\begin{bmatrix} 0 & 0 \\ 0 & R \end{bmatrix}[\ \gamma(L) \quad (1-\gamma(L)\ ]^T\} \\ &= R(1-p) \end{aligned}$$

by taking into account of $E_\gamma[\gamma(L)] = E_\gamma[\gamma^2(L)] = p(L)$.

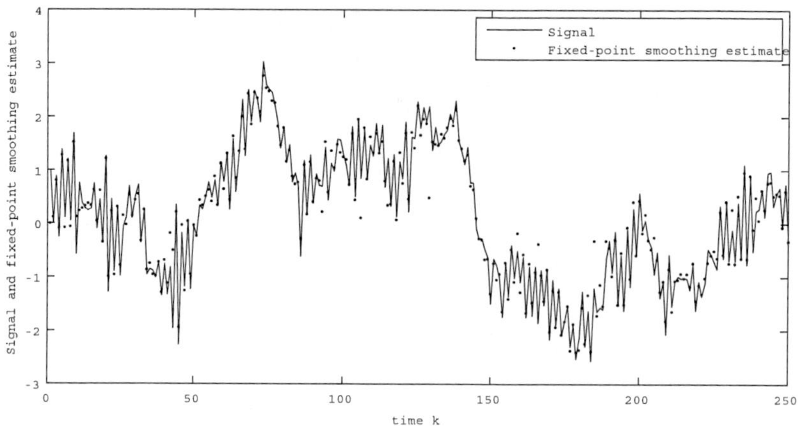

Figure 4.1. Signal $z(k)$ and fixed-point smoothing estimate $\hat{z}(k,k+5)$ vs. k for the white observation noise $N(0,0.1^2)$.

Substituting $\bar{H}$, $\bar{\Phi}$, $K_{\bar{x}}(L,L) = K_{\bar{x}}(0)$, Φ_V, $K_V(L,L)$, $\bar{p}(L) = \begin{bmatrix} p(L) & 1-p(L) \end{bmatrix}$ and $\bar{R}(L)$ into the RLS Wiener estimation algorithms in Theorem 1, the filtering and fixed-point smoothing estimates are calculated recursively. Figure 4.1 illustrates the signal $z(k)$ and the fixed-point smoothing estimate $\hat{z}(k,k+5)$ vs. k for the white observation noise $N(0,0.1^2)$. The probability $p(k)$that the observed value $y(k-1)$ arrives, with one sampling time delay, on time k is 0.039. The probability $1-p(k)$ that the observed value $y(k)$ arrives on time k is 0.961. Figure 4.2 illustrates the mean-square values (MSVs) of the filtering errors $z(k)-\hat{z}(k,k)$ and the fixed-point smoothing errors $z(k)-\hat{z}(k,k+Lag)$ vs. Lag, $0 \leq Lag \leq 5$, for the white Gaussian observation noises $N(0,0.1^2)$, $N(0,0.3^2)$, $N(0,0.5^2)$ and $N(0,0.7^2)$. For $Lag = 0$, the MSV of the filtering errors $z(k)-\hat{z}(k,k)$ is shown. As Lag increases, the MSVs decrease gradually and hence, the estimation accuracy is improved by the fixed-point smoother. For the white Gaussian observation noise with larger variance, the MSVs of the filtering errors and the fixed-point smoothing errors increase and the estimation accuracy worsens. Here, The MSVs of the fixed-point smoothing and filtering errors are evaluated by

$$\sum_{k=1}^{2000}\sum_{j=1}^{Lag}(z(k)-\hat{z}(k,k+j))^2/(2000\times Lag) \text{ and } \sum_{k=1}^{2000}(z(k)-\hat{z}(k,k))^2/2000.$$

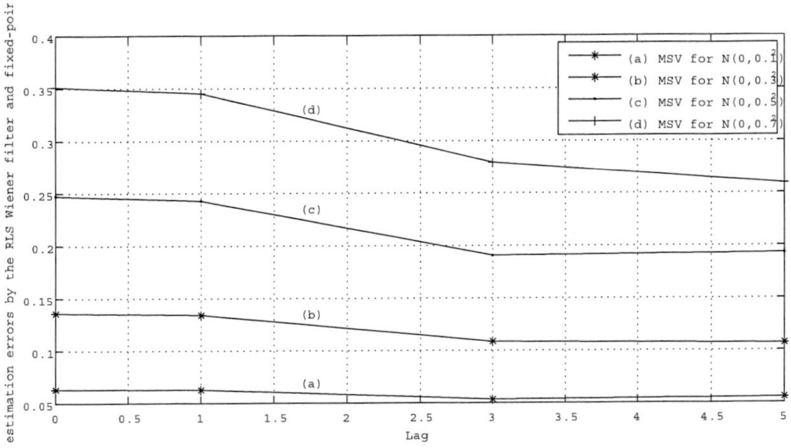

Figure 4.2. MSVs of the filtering errors $z(k)-\hat{z}(k,k)$ and the fixed-point smoothing errors $z(k)-\hat{z}(k,k+Lag)$ vs. Lag, $0 \le Lag \le 5$, for the white Gaussian observation noises $N(0,0.1^2)$, $N(0,0.3^2)$, $N(0,0.5^2)$ and $N(0,0.7^2)$.

Chapter 5

Conclusion

In this monograph, the RLS Wiener filter and fixed-point smoother are designed from randomly delayed observations by one sampling time in linear discrete-time stochastic systems. The probability of the arrival of the observed value $y(k)$ on time k is $1-p(k)$ and the probability of the arrival of the observed value $y(k-1)$ on time k is $p(k)$.

Under the preliminary assumptions of chapter 2, for the observation equation, given by (2.1), in this monograph, by iterative use of the invariant imbedding method, the RLS Wiener algorithms for the filtering estimate $\hat{z}(k,k)$ and the fixed-point smoothing estimate $\hat{z}(k,L)$ of the signal $z(k)$, in addition to $\hat{x}(k,k)$ and $\hat{x}(k,L)$ of the state vector $x(k)$, are derived. The filtering and fixed-point smoothing algorithms take into account the stochastic properties of the random variables $\{\gamma(k)\}$ in the observation equation (2.1) such as the probability $p(k) = P\{\gamma(k) = 1\}$, that the signal exists in the uncertain observation equation, and the $(2,2)$ element $[P(k|j)]_{2,2}$ of the conditional probability of $\gamma(k)$, given $\gamma(j)$. Concerning the simulation in chapter 4, since $\gamma(k)$ is a mixture of two switching sequence, its conditional probability satisfies (2.3) [16].

A numerical simulation example in chapter 4 shows that the fixed-point smoothing and filtering algorithms proposed in this monograph are feasible.

The RLS Wiener estimators do not use the information of the variance of the input noise and the input matrix in the state equation, in comparison with the estimation technique in [3]-[6]. Hence, the RLS Wiener estimation technique has an advantage that the estimation accuracy of the RLS Wiener estimators are not influenced by the estimation errors for the input noise variance and the input matrix.

Appendix

If we subtract the equation obtained by putting $L \to L-1$ in (2.26) from (2.26), we have

$$(h(k,s,L)-h(k,s,L-1))\bar{R}_U(s) = -h(k,L,L)E_\gamma[\bar{\gamma}(L)(\bar{H}K_{\bar{x}}(L,s)\bar{H}^T + \Phi_V K_V(L,s)\Phi_V^T + \Phi_V K_{VU}(L,s) + K_{UV}(L,s)\Phi_V^T)\bar{\gamma}^T(s)] - \sum_{i=1}^{L} h(k,i,L)E_\gamma[\bar{\gamma}(i)(\bar{H}K_{\bar{x}}(i,s)\bar{H}^T + \Phi_V K_V(i,s)\Phi_V^T + \Phi_V K_{VU}(i,s) + K_{UV}(i,s)\Phi_V^T)\bar{\gamma}^T(s)].$$

From (2.24), it is shown that

$$K_{VU}(L,s) = \Phi_V^{L-s}K_{VU}(s,s) + \Phi_V^{L-s-1}R_U(s) = \Phi_V^{L-s-1}R_U(s)\ , \quad K_{VU}(s,s) = 0.$$

Hence,

$$(h(k,s,L)-h(k,s,L-1))\bar{R}_U(s) = -h(k,L,L)E_\gamma[\bar{\gamma}(L)(\bar{H}K_{\bar{x}}(L,s)\bar{H}^T + \Phi_V K_V(L,s)\Phi_V^T + \Phi_V^{L-s}R_U(s))\bar{\gamma}^T(s)] - \sum_{i=1}^{L} h(k,i,L)E_\gamma[\bar{\gamma}(i)(\bar{H}K_{\bar{x}}(i,s)\bar{H}^T + \Phi_V K_V(i,s)\Phi_V^T + \Phi_V K_{VU}(i,s) + K_{UV}(i,s)\Phi_V^T)\bar{\gamma}^T(s)]. \tag{A-1}$$

Let us introduce new functions $J_1(s,L-1)$, $J_2(s,L-1)$ and $J_3(s,L-1)$, which satisfies

$$J_1(s,L-1)\bar{R}_U(s) = \begin{bmatrix} \bar{\Phi}^{-s}K_{\bar{x}}(s,s)\bar{H}^T \\ \Phi_V^{-s}K_V(s,s)\Phi_V^T \\ \Phi_V^{-s}R_U(s) \end{bmatrix} \begin{bmatrix} p(s)I_m \\ 0 \end{bmatrix} - - \sum_{i=1}^{L} J_1(i,L-1)E_\gamma[\bar{\gamma}(i)(\bar{H}K_{\bar{x}}(i,s)\bar{H}^T + \Phi_V K_V(i,s)\Phi_V^T + \Phi_V K_{VU}(i,s) + K_{UV}(i,s)\Phi_V^T)\bar{\gamma}^T(s)], \tag{A-2}$$

$$
\begin{aligned}
J_2(s,L-1)\bar{R}_U(s) &= \begin{bmatrix} \bar{\Phi}^{-s}K_{\bar{x}}(s,s)\bar{H}^T \\ \Phi_V^{-s}K_V(s,s)\Phi_V^T \\ \Phi_V^{-s}R_U(s) \end{bmatrix} \begin{bmatrix} 0 \\ I_m \end{bmatrix} - \\
&- \sum_{i=1}^{L} J_2(i,L-1)E_\gamma[\bar{\gamma}(i)(\bar{H}K_{\bar{x}}(i,s)\bar{H}^T + \Phi_V K_V(i,s)\Phi_V^T + \Phi_V K_{VU}(i,s) \\
&+ K_{UV}(i,s)\Phi_V^T)\bar{\gamma}^T(s)],
\end{aligned} \tag{A-3}
$$

$$
\begin{aligned}
J_3(s,L-1)\bar{R}_U(s) &= \begin{bmatrix} \bar{\Phi}^{-s}K_{\bar{x}}(s,s)\bar{H}^T \\ \Phi_V^{-s}K_V(s,s)\Phi_V^T \\ \Phi_V^{-s}R_U(s) \end{bmatrix} \begin{bmatrix} 0 \\ p(s)I_m \end{bmatrix} - \\
&- \sum_{i=1}^{L} J_3(i,L-1)E_\gamma[\bar{\gamma}(i)(\bar{H}K_{\bar{x}}(i,s)\bar{H}^T + \Phi_V K_V(i,s)\Phi_V^T + \Phi_V K_{VU}(i,s) \\
&+ K_{UV}(i,s)\Phi_V^T)\bar{\gamma}^T(s)].
\end{aligned} \tag{A-4}
$$

From (A-1)-(A-4), we obtain

$$
\begin{aligned}
&h(k,s,L) - h(k,s,L-1) \\
&= -h(k,L,L)\{[\ p_{2,2}(L)I_m \quad 0\ ][\ \bar{H}\bar{\Phi}^L \quad \Phi_V^{L+1} \quad \Phi_V^L\ ]J_1(s,L-1) \\
&+[\ 0 \quad (1-p_{2,2}(L))I_m\ ][\ \bar{H}\bar{\Phi}^L \quad \Phi_V^{L+1} \quad \Phi_V^L\ ]J_1(s,L-1) \\
&+[\ p(L)I_m \quad 0\ ][\ \bar{H}\bar{\Phi}^L \quad \Phi_V^{L+1} \quad \Phi_V^L\ ]J_2(s,L-1) \\
&+[\ 0 \quad (1-p(L))I_m\ ][\ \bar{H}\bar{\Phi}^L \quad \Phi_V^{L+1} \quad \Phi_V^L\ ]J_2(s,L-1) \\
&-[\ p_{2,2}(L)I_m \quad 0\ ][\ \bar{H}\bar{\Phi}^L \quad \Phi_V^{L+1} \quad \Phi_V^L\ ]J_3(s,L-1) \\
&-[\ 0 \quad (1-p_{2,2}(L))I_m\ ][\ \bar{H}\bar{\Phi}^L \quad \Phi_V^{L+1} \quad \Phi_V^L\ ]J_3(s,L-1)\}.
\end{aligned} \tag{A-5}
$$

(A-5) is also written as

$$
\begin{aligned}
&h(k,s,L) - h(k,s,L-1) \\
&= -h(k,L,L)\{[\ p_{2,2}(L)I_m \quad (1-p_{2,2}(L))I_m\ ][\ \bar{H}\bar{\Phi}^L \quad \Phi_V^{L+1} \quad \Phi_V^L\ ]J_1(s,L-1) \\
&+[\ p(L)I_m \quad (1-p(L))I_m 0\ ][\ \bar{H}\bar{\Phi}^L \quad \Phi_V^{L+1} \quad \Phi_V^L\ ]J_2(s,L-1) \\
&-[\ p_{2,2}(L)I_m \quad (1-p_{2,2}(L))I_m\ ][\ \bar{H}\bar{\Phi}^L \quad \Phi_V^{L+1} \quad \Phi_V^L\ ]J_3(s,L-1)\}.
\end{aligned} \tag{A-6}
$$

If we subtract $J_1(s,L-1)\bar{R}(s)$ from $J_1(s,L)\bar{R}(s)$, from (A-2)-(A-4), we obtain

$$
\begin{aligned}
&J_1(s,L) - J_1(s,L-1) = \\
&-J_1(L,L)\{[\ p_{2,2}(L)I_m \quad (1-p_{2,2}(L))I_m\ ][\ \bar{H}\bar{\Phi}^L \quad \Phi_V^{L+1} \quad \Phi_V^L\ ]J_1(s,L-1) \\
&+[\ p(L)I_m \quad (1-p(L))I_m\ ][\ \bar{H}\bar{\Phi}^L \quad \Phi_V^{L+1} \quad \Phi_V^L\ ]J_2(s,L-1) \\
&-[\ p_{2,2}(L)I_m \quad (1-p_{2,2}(L))I_m\ ][\ \bar{H}\bar{\Phi}^L \quad \Phi_V^{L+1} \quad \Phi_V^L\ ]J_3(s,L-1).
\end{aligned}
\tag{A-7}
$$

Similarly, the following equations are obtained.

$$
\begin{aligned}
&J_2(s,L) - J_2(s,L-1) = \\
&-J_2(L,L)\{[\ p_{2,2}(L)I_m \quad (1-p_{2,2}(L))I_m\ ][\ \bar{H}\bar{\Phi}^L \quad \Phi_V^{L+1} \quad \Phi_V^L\ ]J_1(s,L-1) \\
&+[\ p(L)I_m \quad (1-p(L))I_m\ ][\ \bar{H}\bar{\Phi}^L \quad \Phi_V^{L+1} \quad \Phi_V^L\ ]J_2(s,L-1) \\
&-[\ p_{2,2}(L)I_m \quad (1-p_{2,2}(L))I_m\ ][\ \bar{H}\bar{\Phi}^L \quad \Phi_V^{L+1} \quad \Phi_V^L\ ]J_3(s,L-1)
\end{aligned}
\tag{A-8}
$$

$$
\begin{aligned}
&J_3(s,L) - J_3(s,L-1) = \\
&-J_3(L,L)\{[\ p_{2,2}(L)I_m \quad (1-p_{2,2}(L))I_m\ ][\ \bar{H}\bar{\Phi}^L \quad \Phi_V^{L+1} \quad \Phi_V^L\ ]J_1(s,L-1) \\
&+[\ p(L)I_m \quad (1-p(L))I_m\ ][\ \bar{H}\bar{\Phi}^L \quad \Phi_V^{L+1} \quad \Phi_V^L\ ]J_2(s,L-1) \\
&-[\ p_{2,2}(L)I_m \quad (1-p_{2,2}(L))I_m\ ][\ \bar{H}\bar{\Phi}^L \quad \Phi_V^{L+1} \quad \Phi_V^L\ ]J_3(s,L-1)
\end{aligned}
\tag{A-9}
$$

From (A-2), we have an expression for $J_1(L,L)$ as follows.

$$
\begin{aligned}
J_1(L,L)\bar{R}_U(L) = &\begin{bmatrix} \bar{\Phi}^{-L}K_{\bar{x}}(L,L)\bar{H}^T \\ \Phi_V^{-L}K_V(L,L)\Phi_V^T \\ \Phi_V^{-L}R_U(L) \end{bmatrix} \begin{bmatrix} p(L)I_m \\ 0 \end{bmatrix} - \\
&- \sum_{i=1}^{L} J_1(i,L)E_\gamma[\bar{\gamma}(i)(\bar{H}K_{\bar{x}}(i,L)\bar{H}^T + \Phi_V K_V(i,L)\Phi_V^T \\
&+\Phi_V K_{VU}(i,L) + K_{UV}(i,L)\Phi_V^T)\bar{\gamma}^T(L)] \\
= &\begin{bmatrix} \bar{\Phi}^{-L}K_{\bar{x}}(L,L)\bar{H}^T \\ \Phi_V^{-L}K_V(L,L)\Phi_V^T \\ \Phi_V^{-L}R_U(L) \end{bmatrix} \begin{bmatrix} p(L)I_m \\ 0 \end{bmatrix} \\
&-J_1(L,L)E_\gamma[\bar{\gamma}(L)(\bar{H}K_{\bar{x}}(L,L)\bar{H}^T + \Phi_V K_V(L,L)\Phi_V^T)\bar{\gamma}^T(L)] \\
&- \sum_{i=1}^{L-1} J_1(i,L)E_\gamma[\bar{\gamma}(i)(\bar{H}K_{\bar{x}}(i,L)\bar{H}^T + \Phi_V K_V(i,L)\Phi_V^T \\
&+\Phi_V K_{VU}(i,L) + K_{UV}(i,L)\Phi_V^T)\bar{\gamma}^T(L)]
\end{aligned}
\tag{A-10}
$$

From the relationship

$$
\begin{aligned}
&E_\gamma[\bar{\gamma}(L)(\bar{H}K_{\bar{x}}(L,L)\bar{H}^T+\Phi_V K_V(L,L)\Phi_V^T)\bar{\gamma}^T(L)]\\
&=\begin{bmatrix} p(L)I_m & 0 \end{bmatrix}[\bar{H}K_{\bar{x}}(L,L)\bar{H}^T+\Phi_V K_V(L,L)\Phi_V^T]\begin{bmatrix} I_m \\ 0 \end{bmatrix}\\
&+\begin{bmatrix} 0 & (1-p(L))I_m \end{bmatrix}[\bar{H}K_{\bar{x}}(L,L)\bar{H}^T+\Phi_V K_V(L,L)\Phi_V^T]\begin{bmatrix} 0 \\ I_m \end{bmatrix},
\end{aligned}
$$

(A-10) is rewritten, from $K_{VU}(i,L)=0$, as

$$
\begin{aligned}
&J_1(L,L)\Lambda(L)=\begin{bmatrix} \bar{\Phi}^{-L}K_{\bar{x}}(L,L)\bar{H}^T \\ \Phi_V^{-L}K_V(L,L)\Phi_V^T \\ \Phi_V^{-L}R_U(L) \end{bmatrix}\begin{bmatrix} p(L)I_m \\ 0 \end{bmatrix}-\\
&-\sum_{i=1}^{L-1} J_1(i,L)E_\gamma[\bar{\gamma}(i)(\bar{H}K_{\bar{x}}(i,L)\bar{H}^T+\Phi_V K_V(i,L)\Phi_V^T\\
&+\Phi_V K_{VU}(i,L)+K_{UV}(i,L)\Phi_V^T)\bar{\gamma}^T(L)]\\
&=\begin{bmatrix} \bar{\Phi}^{-L}K_{\bar{x}}(L,L)\bar{H}^T \\ \Phi_V^{-L}K_V(L,L)\Phi_V^T \\ \Phi_V^{-L}R_U(L) \end{bmatrix}\begin{bmatrix} p(L)I_m \\ 0 \end{bmatrix}-\\
&-\sum_{i=1}^{L-1} J_1(i,L)E_\gamma[\bar{\gamma}(i)(\bar{H}K_{\bar{x}}(i,i)(\bar{\Phi}^T)^{-i}(\bar{\Phi}^T)^L\bar{H}^T\\
&+\Phi_V K_V(i,i)(\Phi_V^T)^{-i}(\Phi_V^T)^{L+1}+R_U(i)(\Phi_V^T)^{-i}(\Phi_V^T)^L)\bar{\gamma}^T(L)].
\end{aligned}
\tag{A-11}
$$

Here, $\Lambda(L)$ is given by (3.23). Taking into account of the relationship

$$
\begin{aligned}
&E_{\gamma}[\bar{\gamma}(i)(\bar{H}K_{\bar{x}}(i,i)(\bar{\Phi}^T)^{-i}(\bar{\Phi}^T)^L\bar{H}^T + \Phi_V K_V(i,i)(\Phi_V^T)^{-i}(\Phi_V^T)^{L+1} \\
&+R_U(i)(\Phi_V^T)^{-i}(\Phi_V^T)^L)\bar{\gamma}^T(L)] \\
&= [\ p(i)I_m \quad 0\ ][\bar{H}K_{\bar{x}}(i,i)(\bar{\Phi}^T)^{-i}(\bar{\Phi}^T)^L\bar{H}^T + \Phi_V K_V(i,i)(\Phi_V^T)^{-i}(\Phi_V^T)^{L+1} \\
&+R_U(i)(\Phi_V^T)^{-i}(\Phi_V^T)^L] \begin{bmatrix} p_{2,2}(L)I_m \\ 0 \end{bmatrix} \\
&+[\ p(i)I_m \quad 0\ ][\bar{H}K_{\bar{x}}(i,i)(\bar{\Phi}^T)^{-i}(\bar{\Phi}^T)^L\bar{H}^T + \Phi_V K_V(i,i)(\Phi_V^T)^{-i}(\Phi_V^T)^{L+1} \\
&+R_U(i)(\Phi_V^T)^{-i}(\Phi_V^T)^L] \begin{bmatrix} 0 \\ (1-p_{2,2}(L))I_m \end{bmatrix} \\
&+[\ 0 \quad I_m\ ][\bar{H}K_{\bar{x}}(i,i)(\bar{\Phi}^T)^{-i}(\bar{\Phi}^T)^L\bar{H}^T + \Phi_V K_V(i,i)(\Phi_V^T)^{-i}(\Phi_V^T)^{L+1} \\
&+R_U(i)(\Phi_V^T)^{-i}(\Phi_V^T)^L] \begin{bmatrix} p(L)I_m \\ 0 \end{bmatrix} \\
&-[\ 0 \quad p(i)I_m\ ][\bar{H}K_{\bar{x}}(i,i)(\bar{\Phi}^T)^{-i}(\bar{\Phi}^T)^L\bar{H}^T + \Phi_V K_V(i,i)(\Phi_V^T)^{-i}(\Phi_V^T)^{L+1} \\
&+R_U(i)(\Phi_V^T)^{-i}(\Phi_V^T)^L] \begin{bmatrix} p_{2,2}(L)I_m \\ 0 \end{bmatrix} \\
&+[\ 0 \quad (1-p(i))I_m\ ][\bar{H}K_{\bar{x}}(i,i)(\bar{\Phi}^T)^{-i}(\bar{\Phi}^T)^L\bar{H}^T + \Phi_V K_V(i,i)(\Phi_V^T)^{-i}(\Phi_V^T)^{L+1} \\
&+R_U(i)(\Phi_V^T)^{-i}(\Phi_V^T)^L] \begin{bmatrix} 0 \\ I_m \end{bmatrix} \\
&-[\ 0 \quad I_m\ ][\bar{H}K_{\bar{x}}(i,i)(\bar{\Phi}^T)^{-i}(\bar{\Phi}^T)^L\bar{H}^T + \Phi_V K_V(i,i)(\Phi_V^T)^{-i}(\Phi_V^T)^{L+1} \\
&+R_U(i)(\Phi_V^T)^{-i}(\Phi_V^T)^L] \begin{bmatrix} 0 \\ p(L)I_m \end{bmatrix} \\
&+[\ 0 \quad p(i)I_m\ ][\bar{H}K_{\bar{x}}(i,i)(\bar{\Phi}^T)^{-i}(\bar{\Phi}^T)^L\bar{H}^T + \Phi_V K_V(i,i)(\Phi_V^T)^{-i}(\Phi_V^T)^{L+1} \\
&+R_U(i)(\Phi_V^T)^{-i}(\Phi_V^T)^L] \begin{bmatrix} p_{2,2}(L)I_m \\ 0 \end{bmatrix},
\end{aligned}
$$

and introducing the functions

$$
\begin{aligned}
r_{11}(L-1) = \sum_{i=1}^{L-1} J_1(i,L-1) \\
[\ p(i)I_m \quad 0\ ][\ \bar{H}K_{\bar{x}}(i,i)(\bar{\Phi}^T)^{-i} \quad \Phi_V K_V(i,i)(\Phi_V^T)^{-i} \quad R_U(i)(\Phi_V^T)^{-i}\ ],
\end{aligned}
$$

$$
\begin{aligned}
r_{12}(L-1) = \sum_{i=1}^{L-1} J_1(i,L-1) \\
[\ 0 \quad I_m\ ][\ \bar{H}K_{\bar{x}}(i,i)(\bar{\Phi}^T)^{-i} \quad \Phi_V K_V(i,i)(\Phi_V^T)^{-i} \quad R_U(i)(\Phi_V^T)^{-i}\ ],
\end{aligned}
$$

$$r_{13}(L-1) = \sum_{i=1}^{L-1} J_1(i, L-1)$$
$$[\ 0 \quad p(i)I_m\ ][\ \bar{H}K_{\bar{x}}(i,i)(\bar{\Phi}^T)^{-i} \quad \Phi_V K_V(i,i)(\Phi_V^T)^{-i} \quad R_U(i)(\Phi_V^T)^{-i}\ ],$$

$$r_{21}(L-1) = \sum_{i=1}^{L-1} J_2(i, L-1)$$
$$[\ p(i)I_m \quad 0\ ][\ \bar{H}K_{\bar{x}}(i,i)(\bar{\Phi}^T)^{-i} \quad \Phi_V K_V(i,i)(\Phi_V^T)^{-i} \quad R_U(i)(\Phi_V^T)^{-i}\ ],$$

$$r_{22}(L-1) = \sum_{i=1}^{L-1} J_2(i, L-1)$$
$$[\ 0 \quad I_m\ ][\ \bar{H}K_{\bar{x}}(i,i)(\bar{\Phi}^T)^{-i} \quad \Phi_V K_V(i,i)(\Phi_V^T)^{-i} \quad R_U(i)(\Phi_V^T)^{-i}\ ],$$

$$r_{23}(L-1) = \sum_{i=1}^{L-1} J_2(i, L-1)$$
$$[\ 0 \quad p(i)I_m\ ][\ \bar{H}K_{\bar{x}}(i,i)(\bar{\Phi}^T)^{-i} \quad \Phi_V K_V(i,i)(\Phi_V^T)^{-i} \quad R_U(i)(\Phi_V^T)^{-i}\ ],$$

$$r_{31}(L-1) = \sum_{i=1}^{L-1} J_3(i, L-1)$$
$$[\ p(i)I_m \quad 0\ ][\ \bar{H}K_{\bar{x}}(i,i)(\bar{\Phi}^T)^{-i} \quad \Phi_V K_V(i,i)(\Phi_V^T)^{-i} \quad R_U(i)(\Phi_V^T)^{-i}\ ],$$

$$r_{32}(L-1) = \sum_{i=1}^{L-1} J_3(i, L-1)$$
$$[\ 0 \quad I_m\ ][\ \bar{H}K_{\bar{x}}(i,i)(\bar{\Phi}^T)^{-i} \quad \Phi_V K_V(i,i)(\Phi_V^T)^{-i} \quad R_U(i)(\Phi_V^T)^{-i}\ ],$$

$$r_{33}(L-1) = \sum_{i=1}^{L-1} J_3(i, L-1)$$
$$[\ 0 \quad p(i)I_m\ ][\ \bar{H}K_{\bar{x}}(i,i)(\bar{\Phi}^T)^{-i} \quad \Phi_V K_V(i,i)(\Phi_V^T)^{-i} \quad R_U(i)(\Phi_V^T)^{-i}\ ],$$

we have an equation for $J_1(L,L)$ as

$$J_1(L,L) = \{ \begin{bmatrix} \bar{\Phi}^{-L} K_{\bar{x}}(L,L) \bar{H}^T \\ \Phi_V^{-L} K_V(L,L) \Phi_V^T \\ \Phi_V^{-L} R_U(L) \end{bmatrix} \begin{bmatrix} p(L) I_m \\ 0 \end{bmatrix}$$
$$-r_{11}(L-1) \begin{bmatrix} (\bar{\Phi}^T)\bar{H}^T \\ (\Phi_V^T)^{L+1} \\ (\Phi_V^T)^L \end{bmatrix} \begin{bmatrix} p_{2,2}(L) I_m \\ (1-p_{2,2}(L)) I_m \end{bmatrix}$$
$$-r_{12}(L-1) \begin{bmatrix} (\bar{\Phi}^T)\bar{H}^T \\ (\Phi_V^T)^{L+1} \\ (\Phi_V^T)^L \end{bmatrix} \begin{bmatrix} p(L) I_m \\ (1-p(L)) I_m \end{bmatrix} \tag{A-12}$$
$$+r_{13}(L-1) \begin{bmatrix} (\bar{\Phi}^T)\bar{H}^T \\ (\Phi_V^T)^{L+1} \\ (\Phi_V^T)^L \end{bmatrix} \begin{bmatrix} p_{2,2}(L) I_m \\ (1-p_{2,2}(L)) I_m \end{bmatrix} \} \Gamma^{-1}(L).$$

Here, $\Gamma(L)$ is given by (3.22) and the following functions have been introduced.

$$S_{1111}(L) = \bar{\Phi}^L r_{1111}(L)(\bar{\Phi}^T)^L, \quad S_{1112}(L) = \bar{\Phi}^L r_{1112}(L)(\bar{\Phi}^T)^L, \quad S_{1113}(L) = \bar{\Phi}^L r_{1113}(L)(\bar{\Phi}^T)^L,$$
$$S_{1121}(L) = \bar{\Phi}^L r_{1121}(L)(\bar{\Phi}^T)^L, \quad S_{1122}(L) = \bar{\Phi}^L r_{1122}(L)(\bar{\Phi}^T)^L, \quad S_{1123}(L) = \bar{\Phi}^L r_{1123}(L)(\bar{\Phi}^T)^L,$$
$$S_{1131}(L) = \bar{\Phi}^L r_{1131}(L)(\bar{\Phi}^T)^L, \quad S_{1132}(L) = \bar{\Phi}^L r_{1132}(L)(\bar{\Phi}^T)^L, \quad S_{1133}(L) = \bar{\Phi}^L r_{1133}(L)(\bar{\Phi}^T)^L,$$

$$S_{1211}(L) = \bar{\Phi}^L r_{1211}(L)(\bar{\Phi}^T)^L, \quad S_{1212}(L) = \bar{\Phi}^L r_{1212}(L)(\bar{\Phi}^T)^L, \quad S_{1213}(L) = \bar{\Phi}^L r_{1213}(L)(\bar{\Phi}^T)^L,$$
$$S_{1221}(L) = \bar{\Phi}^L r_{1221}(L)(\bar{\Phi}^T)^L, \quad S_{1222}(L) = \bar{\Phi}^L r_{1222}(L)(\bar{\Phi}^T)^L, \quad S_{1223}(L) = \bar{\Phi}^L r_{1223}(L)(\bar{\Phi}^T)^L,$$
$$S_{1231}(L) = \bar{\Phi}^L r_{1231}(L)(\bar{\Phi}^T)^L, \quad S_{1232}(L) = \bar{\Phi}^L r_{1232}(L)(\bar{\Phi}^T)^L, \quad S_{1233}(L) = \bar{\Phi}^L r_{1233}(L)(\bar{\Phi}^T)^L,$$

$$S_{1311}(L) = \bar{\Phi}^L r_{1311}(L)(\bar{\Phi}^T)^L, \quad S_{1312}(L) = \bar{\Phi}^L r_{1312}(L)(\bar{\Phi}^T)^L, \quad S_{1313}(L) = \bar{\Phi}^L r_{1313}(L)(\bar{\Phi}^T)^L,$$
$$S_{1321}(L) = \bar{\Phi}^L r_{1321}(L)(\bar{\Phi}^T)^L, \quad S_{1322}(L) = \bar{\Phi}^L r_{1322}(L)(\bar{\Phi}^T)^L, \quad S_{1323}(L) = \bar{\Phi}^L r_{1323}(L)(\bar{\Phi}^T)^L,$$
$$S_{1331}(L) = \bar{\Phi}^L r_{1331}(L)(\bar{\Phi}^T)^L, \quad S_{1332}(L) = \bar{\Phi}^L r_{1332}(L)(\bar{\Phi}^T)^L, \quad S_{1333}(L) = \bar{\Phi}^L r_{1333}(L)(\bar{\Phi}^T)^L,$$

$$S_{2111}(L) = \bar{\Phi}^L r_{2111}(L)(\bar{\Phi}^T)^L, \quad S_{2112}(L) = \bar{\Phi}^L r_{2112}(L)(\bar{\Phi}^T)^L, \quad S_{2113}(L) = \bar{\Phi}^L r_{2113}(L)(\bar{\Phi}^T)^L,$$
$$S_{2121}(L) = \bar{\Phi}^L r_{2121}(L)(\bar{\Phi}^T)^L, \quad S_{2122}(L) = \bar{\Phi}^L r_{2122}(L)(\bar{\Phi}^T)^L, \quad S_{2123}(L) = \bar{\Phi}^L r_{2123}(L)(\bar{\Phi}^T)^L,$$
$$S_{2131}(L) = \bar{\Phi}^L r_{2131}(L)(\bar{\Phi}^T)^L, \quad S_{2132}(L) = \bar{\Phi}^L r_{2132}(L)(\bar{\Phi}^T)^L, \quad S_{2133}(L) = \bar{\Phi}^L r_{2133}(L)(\bar{\Phi}^T)^L,$$

$$S_{2211}(L) = \bar{\Phi}^L r_{2211}(L)(\bar{\Phi}^T)^L, \quad S_{2212}(L) = \bar{\Phi}^L r_{2212}(L)(\bar{\Phi}^T)^L, \quad S_{2213}(L) = \bar{\Phi}^L r_{2213}(L)(\bar{\Phi}^T)^L,$$
$$S_{2221}(L) = \bar{\Phi}^L r_{2221}(L)(\bar{\Phi}^T)^L, \quad S_{2222}(L) = \bar{\Phi}^L r_{2222}(L)(\bar{\Phi}^T)^L, \quad S_{2223}(L) = \bar{\Phi}^L r_{2223}(L)(\bar{\Phi}^T)^L,$$
$$S_{2231}(L) = \bar{\Phi}^L r_{2231}(L)(\bar{\Phi}^T)^L, \quad S_{2232}(L) = \bar{\Phi}^L r_{2232}(L)(\bar{\Phi}^T)^L, \quad S_{2233}(L) = \bar{\Phi}^L r_{2233}(L)(\bar{\Phi}^T)^L,$$

$$S_{2311}(L) = \bar{\Phi}^L r_{2311}(L)(\bar{\Phi}^T)^L, \quad S_{2312}(L) = \bar{\Phi}^L r_{2312}(L)(\bar{\Phi}^T)^L, \quad S_{2313}(L) = \bar{\Phi}^L r_{2313}(L)(\bar{\Phi}^T)^L,$$
$$S_{2321}(L) = \bar{\Phi}^L r_{2321}(L)(\bar{\Phi}^T)^L, \quad S_{2322}(L) = \bar{\Phi}^L r_{2322}(L)(\bar{\Phi}^T)^L, \quad S_{2323}(L) = \bar{\Phi}^L r_{2323}(L)(\bar{\Phi}^T)^L,$$
$$S_{2331}(L) = \bar{\Phi}^L r_{2331}(L)(\bar{\Phi}^T)^L, \quad S_{2332}(L) = \bar{\Phi}^L r_{2332}(L)(\bar{\Phi}^T)^L, \quad S_{2333}(L) = \bar{\Phi}^L r_{2333}(L)(\bar{\Phi}^T)^L,$$

$$S_{3111}(L) = \bar{\Phi}^L r_{3111}(L)(\bar{\Phi}^T)^L, \quad S_{3112}(L) = \bar{\Phi}^L r_{3112}(L)(\bar{\Phi}^T)^L, \quad S_{3113}(L) = \bar{\Phi}^L r_{3113}(L)(\bar{\Phi}^T)^L,$$
$$S_{3121}(L) = \bar{\Phi}^L r_{3121}(L)(\bar{\Phi}^T)^L, \quad S_{3122}(L) = \bar{\Phi}^L r_{3122}(L)(\bar{\Phi}^T)^L, \quad S_{3123}(L) = \bar{\Phi}^L r_{3123}(L)(\bar{\Phi}^T)^L,$$
$$S_{3131}(L) = \bar{\Phi}^L r_{3131}(L)(\bar{\Phi}^T)^L, \quad S_{3132}(L) = \bar{\Phi}^L r_{3132}(L)(\bar{\Phi}^T)^L, \quad S_{3133}(L) = \bar{\Phi}^L r_{3133}(L)(\bar{\Phi}^T)^L,$$

$$S_{3211}(L) = \bar{\Phi}^L r_{3211}(L)(\bar{\Phi}^T)^L, \quad S_{3212}(L) = \bar{\Phi}^L r_{3212}(L)(\bar{\Phi}^T)^L, \quad S_{3213}(L) = \bar{\Phi}^L r_{3213}(L)(\bar{\Phi}^T)^L,$$
$$S_{3221}(L) = \bar{\Phi}^L r_{3221}(L)(\bar{\Phi}^T)^L, \quad S_{3222}(L) = \bar{\Phi}^L r_{3222}(L)(\bar{\Phi}^T)^L, \quad S_{3223}(L) = \bar{\Phi}^L r_{3223}(L)(\bar{\Phi}^T)^L,$$
$$S_{3231}(L) = \bar{\Phi}^L r_{3231}(L)(\bar{\Phi}^T)^L, \quad S_{3232}(L) = \bar{\Phi}^L r_{3232}(L)(\bar{\Phi}^T)^L, \quad S_{3233}(L) = \bar{\Phi}^L r_{3233}(L)(\bar{\Phi}^T)^L,$$

$$S_{3311}(L) = \bar{\Phi}^L r_{3311}(L)(\bar{\Phi}^T)^L, \quad S_{3312}(L) = \bar{\Phi}^L r_{3312}(L)(\bar{\Phi}^T)^L, \quad S_{3313}(L) = \bar{\Phi}^L r_{3313}(L)(\bar{\Phi}^T)^L,$$
$$S_{3321}(L) = \bar{\Phi}^L r_{3321}(L)(\bar{\Phi}^T)^L, \quad S_{3322}(L) = \bar{\Phi}^L r_{3322}(L)(\bar{\Phi}^T)^L, \quad S_{3323}(L) = \bar{\Phi}^L r_{3323}(L)(\bar{\Phi}^T)^L,$$
$$S_{3331}(L) = \bar{\Phi}^L r_{3331}(L)(\bar{\Phi}^T)^L, \quad S_{3332}(L) = \bar{\Phi}^L r_{3332}(L)(\bar{\Phi}^T)^L, \quad S_{3333}(L) = \bar{\Phi}^L r_{3333}(L)(\bar{\Phi}^T)^L,$$

$$r_{1111}(L) = \sum_{i=1}^{L} J_{11}(i,L)[\ p(i)I_m \quad 0\ ]\bar{H}K_{\bar{x}}(i,i)(\bar{\Phi})^{-i},$$

$$r_{1112}(L) = \sum_{i=1}^{L} J_{11}(i,L)[\ p(i)I_m \quad 0\ ]\Phi_V K_V(i,i)(\Phi_V^T)^{-i},$$

$$r_{1113}(L) = \sum_{i=1}^{L} J_{11}(i,L)[\ p(i)I_m \quad 0\ ]R_U(i)(\Phi_V^T)^{-i},$$

$$r_{1121}(L) = \sum_{i=1}^{L} J_{12}(i,L)[\ p(i)I_m \quad 0\ ]\bar{H}K_{\bar{x}}(i,i)(\bar{\Phi})^{-i},$$

$$r_{1122}(L) = \sum_{i=1}^{L} J_{12}(i,L)[\ p(i)I_m \quad 0\ ]\Phi_V K_V(i,i)(\Phi_V^T)^{-i},$$

$$r_{1123}(L) = \sum_{i=1}^{L} J_{12}(i,L)[\ p(i)I_m \quad 0\ ]R_U(i)(\Phi_V^T)^{-i},$$

$$r_{1131}(L) = \sum_{i=1}^{L} J_{13}(i,L)[\ p(i)I_m \quad 0\ ]\bar{H}K_{\bar{x}}(i,i)(\bar{\Phi})^{-i},$$

$$r_{1132}(L) = \sum_{i=1}^{L} J_{13}(i,L)[\ p(i)I_m \quad 0\ ]\Phi_V K_V(i,i)(\Phi_V^T)^{-i},$$

$$r_{1133}(L) = \sum_{i=1}^{L} J_{13}(i,L)[\ p(i)I_m \quad 0\ ]R_U(i)(\Phi_V^T)^{-i},$$

$$r_{1211}(L) = \sum_{i=1}^{L} J_{11}(i,L)[\ 0 \quad I_m\ ]\bar{H}K_{\bar{x}}(i,i)(\bar{\Phi})^{-i},$$

$$r_{1212}(L) = \sum_{i=1}^{L} J_{11}(i,L)[\ 0 \quad I_m\ ]\Phi_V K_V(i,i)(\Phi_V^T)^{-i},$$

$$r_{1213}(L) = \sum_{i=1}^{L} J_{11}(i,L)[\ 0 \quad I_m\ ]R_U(i)(\Phi_V^T)^{-i},$$

$$r_{1221}(L) = \sum_{i=1}^{L} J_{12}(i,L)[\ 0 \quad I_m\ ]\bar{H}K_{\bar{x}}(i,i)(\bar{\Phi})^{-i},$$

$$r_{1222}(L) = \sum_{i=1}^{L} J_{12}(i,L)[\ 0 \quad I_m\ ]\Phi_V K_V(i,i)(\Phi_V^T)^{-i},$$

$$r_{1223}(L) = \sum_{i=1}^{L} J_{12}(i,L)[\ 0 \quad I_m\ ]R_U(i)(\Phi_V^T)^{-i},$$

$$r_{1231}(L) = \sum_{i=1}^{L} J_{13}(i,L)[\ 0 \quad I_m\ ]\bar{H}K_{\bar{x}}(i,i)(\bar{\Phi})^{-i},$$

$$r_{1232}(L) = \sum_{i=1}^{L} J_{13}(i,L)[\ 0 \quad I_m\ ]\Phi_V K_V(i,i)(\Phi_V^T)^{-i},$$

$$r_{1233}(L) = \sum_{i=1}^{L} J_{13}(i,L)[\ 0 \quad I_m\ ]R_U(i)(\Phi_V^T)^{-i},$$

$$r_{1311}(L) = \sum_{i=1}^{L} J_{11}(i,L)[\ 0 \quad p(i)I_m\ ]\bar{H}K_{\bar{x}}(i,i)(\bar{\Phi})^{-i},$$

$$r_{1312}(L) = \sum_{i=1}^{L} J_{11}(i,L)[\ 0 \quad p(i)I_m\ ]\Phi_V K_V(i,i)(\Phi_V^T)^{-i},$$

$$r_{1313}(L) = \sum_{i=1}^{L} J_{11}(i,L)[\ 0 \quad p(i)I_m\ ]R_U(i)(\Phi_V^T)^{-i},$$

$$r_{1321}(L) = \sum_{i=1}^{L} J_{12}(i,L)[\ 0 \quad p(i)I_m\ ]\bar{H}K_{\bar{x}}(i,i)(\bar{\Phi})^{-i},$$

$$r_{1322}(L) = \sum_{i=1}^{L} J_{12}(i,L)[\ 0 \quad p(i)I_m\ ]\Phi_V K_V(i,i)(\Phi_V^T)^{-i},$$

$$r_{1323}(L) = \sum_{i=1}^{L} J_{12}(i,L)[\ 0 \quad p(i)I_m\ ]R_U(i)(\Phi_V^T)^{-i},$$

$$r_{1331}(L) = \sum_{i=1}^{L} J_{13}(i,L)[\ 0 \quad p(i)I_m\ ]\bar{H}K_{\bar{x}}(i,i)(\bar{\Phi})^{-i},$$

$$r_{1332}(L) = \sum_{i=1}^{L} J_{13}(i,L)[\ 0 \quad p(i)I_m\ ]\Phi_V K_V(i,i)(\Phi_V^T)^{-i},$$

$$r_{1333}(L) = \sum_{i=1}^{L} J_{13}(i,L)[\ 0 \quad p(i)I_m\ ]R_U(i)(\Phi_V^T)^{-i},$$

$$r_{2111}(L) = \sum_{i=1}^{L} J_{21}(i,L)[\ p(i)I_m \quad 0\ ]\bar{H}K_{\bar{x}}(i,i)(\bar{\Phi})^{-i},$$

$$r_{2112}(L) = \sum_{i=1}^{L} J_{21}(i,L)[\ p(i)I_m \quad 0\ ]\Phi_V K_V(i,i)(\Phi_V^T)^{-i},$$

$$r_{2113}(L) = \sum_{i=1}^{L} J_{21}(i,L)[\ p(i)I_m \quad 0\ ]R_U(i)(\Phi_V^T)^{-i},$$

$$r_{2121}(L)=\sum_{i=1}^{L}J_{22}(i,L)[\;p(i)I_m\quad 0\;]\bar{H}K_{\bar{x}}(i,i)(\bar{\Phi})^{-i},$$

$$r_{2122}(L)=\sum_{i=1}^{L}J_{22}(i,L)[\;p(i)I_m\quad 0\;]\Phi_V K_V(i,i)(\Phi_V^T)^{-i},$$

$$r_{2123}(L)=\sum_{i=1}^{L}J_{22}(i,L)[\;p(i)I_m\quad 0\;]R_U(i)(\Phi_V^T)^{-i},$$

$$r_{2131}(L)=\sum_{i=1}^{L}J_{23}(i,L)[\;p(i)I_m\quad 0\;]\bar{H}K_{\bar{x}}(i,i)(\bar{\Phi})^{-i},$$

$$r_{2132}(L)=\sum_{i=1}^{L}J_{23}(i,L)[\;p(i)I_m\quad 0\;]\Phi_V K_V(i,i)(\Phi_V^T)^{-i},$$

$$r_{2133}(L)=\sum_{i=1}^{L}J_{23}(i,L)[\;p(i)I_m\quad 0\;]R_U(i)(\Phi_V^T)^{-i},$$

$$r_{2211}(L)=\sum_{i=1}^{L}J_{21}(i,L)[\;0\quad I_m\;]\bar{H}K_{\bar{x}}(i,i)(\bar{\Phi})^{-i},$$

$$r_{2212}(L)=\sum_{i=1}^{L}J_{21}(i,L)[\;0\quad I_m\;]\Phi_V K_V(i,i)(\Phi_V^T)^{-i},$$

$$r_{2213}(L)=\sum_{i=1}^{L}J_{21}(i,L)[\;0\quad I_m\;]R_U(i)(\Phi_V^T)^{-i},$$

$$r_{2221}(L)=\sum_{i=1}^{L}J_{22}(i,L)[\;0\quad I_m\;]\bar{H}K_{\bar{x}}(i,i)(\bar{\Phi})^{-i},$$

$$r_{2222}(L)=\sum_{i=1}^{L}J_{22}(i,L)[\;0\quad I_m\;]\Phi_V K_V(i,i)(\Phi_V^T)^{-i},$$

$$r_{2223}(L)=\sum_{i=1}^{L}J_{22}(i,L)[\;0\quad I_m\;]R_U(i)(\Phi_V^T)^{-i},$$

$$r_{2231}(L)=\sum_{i=1}^{L}J_{23}(i,L)[\;0\quad I_m\;]\bar{H}K_{\bar{x}}(i,i)(\bar{\Phi})^{-i},$$

$$r_{2232}(L)=\sum_{i=1}^{L}J_{23}(i,L)[\;0\quad I_m\;]\Phi_V K_V(i,i)(\Phi_V^T)^{-i},$$

$$r_{2233}(L)=\sum_{i=1}^{L}J_{23}(i,L)[\;0\quad I_m\;]R_U(i)(\Phi_V^T)^{-i},$$

$$r_{2311}(L)=\sum_{i=1}^{L}J_{21}(i,L)[\;0\quad p(i)I_m\;]\bar{H}K_{\bar{x}}(i,i)(\bar{\Phi})^{-i},$$

$$r_{2312}(L)=\sum_{i=1}^{L}J_{21}(i,L)[\ 0\quad p(i)I_m\ ]\Phi_V K_V(i,i)(\Phi_V^T)^{-i},$$

$$r_{2313}(L)=\sum_{i=1}^{L}J_{21}(i,L)[\ 0\quad p(i)I_m\ ]R_U(i)(\Phi_V^T)^{-i},$$

$$r_{2321}(L)=\sum_{i=1}^{L}J_{22}(i,L)[\ 0\quad p(i)I_m\ ]\bar{H}K_{\bar{x}}(i,i)(\bar{\Phi})^{-i},$$

$$r_{2322}(L)=\sum_{i=1}^{L}J_{22}(i,L)[\ 0\quad p(i)I_m\ ]\Phi_V K_V(i,i)(\Phi_V^T)^{-i},$$

$$r_{2323}(L)=\sum_{i=1}^{L}J_{22}(i,L)[\ 0\quad p(i)I_m\ ]R_U(i)(\Phi_V^T)^{-i},$$

$$r_{2331}(L)=\sum_{i=1}^{L}J_{23}(i,L)[\ 0\quad p(i)I_m\ ]\bar{H}K_{\bar{x}}(i,i)(\bar{\Phi})^{-i},$$

$$r_{2332}(L)=\sum_{i=1}^{L}J_{23}(i,L)[\ 0\quad p(i)I_m\ ]\Phi_V K_V(i,i)(\Phi_V^T)^{-i},$$

$$r_{2333}(L)=\sum_{i=1}^{L}J_{23}(i,L)[\ 0\quad p(i)I_m\ ]R_U(i)(\Phi_V^T)^{-i},$$

$$r_{3111}(L)=\sum_{i=1}^{L}J_{31}(i,L)[\ p(i)I_m\quad 0\ ]\bar{H}K_{\bar{x}}(i,i)(\bar{\Phi})^{-i},$$

$$r_{3112}(L)=\sum_{i=1}^{L}J_{31}(i,L)[\ p(i)I_m\quad 0\ ]\Phi_V K_V(i,i)(\Phi_V^T)^{-i},$$

$$r_{3113}(L)=\sum_{i=1}^{L}J_{31}(i,L)[\ p(i)I_m\quad 0\ ]R_U(i)(\Phi_V^T)^{-i},$$

$$r_{3121}(L)=\sum_{i=1}^{L}J_{32}(i,L)[\ p(i)I_m\quad 0\ ]\bar{H}K_{\bar{x}}(i,i)(\bar{\Phi})^{-i},$$

$$r_{3122}(L)=\sum_{i=1}^{L}J_{32}(i,L)[\ p(i)I_m\quad 0\ ]\Phi_V K_V(i,i)(\Phi_V^T)^{-i},$$

$$r_{3123}(L)=\sum_{i=1}^{L}J_{32}(i,L)[\ p(i)I_m\quad 0\ ]R_U(i)(\Phi_V^T)^{-i},$$

$$r_{3131}(L)=\sum_{i=1}^{L}J_{33}(i,L)[\ p(i)I_m\quad 0\ ]\bar{H}K_{\bar{x}}(i,i)(\bar{\Phi})^{-i},$$

$$r_{3132}(L)=\sum_{i=1}^{L}J_{33}(i,L)[\ p(i)I_m\quad 0\ ]\Phi_V K_V(i,i)(\Phi_V^T)^{-i},$$

$$r_{3133}(L) = \sum_{i=1}^{L} J_{33}(i,L)[\ p(i)I_m \quad 0\ ]R_U(i)(\Phi_V^T)^{-i},$$

$$r_{3211}(L) = \sum_{i=1}^{L} J_{31}(i,L)[\ 0 \quad I_m\ ]\bar{H}K_{\bar{x}}(i,i)(\bar{\Phi})^{-i},$$

$$r_{3212}(L) = \sum_{i=1}^{L} J_{31}(i,L)[\ 0 \quad I_m\ ]\Phi_V K_V(i,i)(\Phi_V^T)^{-i},$$

$$r_{3213}(L) = \sum_{i=1}^{L} J_{31}(i,L)[\ 0 \quad I_m\ ]R_U(i)(\Phi_V^T)^{-i},$$

$$r_{3221}(L) = \sum_{i=1}^{L} J_{32}(i,L)[\ 0 \quad I_m\ ]\bar{H}K_{\bar{x}}(i,i)(\bar{\Phi})^{-i},$$

$$r_{3222}(L) = \sum_{i=1}^{L} J_{32}(i,L)[\ 0 \quad I_m\ ]\Phi_V K_V(i,i)(\Phi_V^T)^{-i},$$

$$r_{3223}(L) = \sum_{i=1}^{L} J_{32}(i,L)[\ 0 \quad I_m\ ]R_U(i)(\Phi_V^T)^{-i},$$

$$r_{3231}(L) = \sum_{i=1}^{L} J_{33}(i,L)[\ 0 \quad I_m\ ]\bar{H}K_{\bar{x}}(i,i)(\bar{\Phi})^{-i},$$

$$r_{3232}(L) = \sum_{i=1}^{L} J_{33}(i,L)[\ 0 \quad I_m\ ]\Phi_V K_V(i,i)(\Phi_V^T)^{-i},$$

$$r_{3233}(L) = \sum_{i=1}^{L} J_{33}(i,L)[\ 0 \quad I_m\ ]R_U(i)(\Phi_V^T)^{-i},$$

$$r_{3311}(L) = \sum_{i=1}^{L} J_{11}(i,L)[\ 0 \quad p(i)I_m\ ]\bar{H}K_{\bar{x}}(i,i)(\bar{\Phi})^{-i},$$

$$r_{3312}(L) = \sum_{i=1}^{L} J_{11}(i,L)[\ 0 \quad p(i)I_m\ ]\Phi_V K_V(i,i)(\Phi_V^T)^{-i},$$

$$r_{3313}(L) = \sum_{i=1}^{L} J_{11}(i,L)[\ 0 \quad p(i)I_m\ ]R_U(i)(\Phi_V^T)^{-i},$$

$$r_{3321}(L) = \sum_{i=1}^{L} J_{12}(i,L)[\ 0 \quad p(i)I_m\ ]\bar{H}K_{\bar{x}}(i,i)(\bar{\Phi})^{-i},$$

$$r_{3322}(L) = \sum_{i=1}^{L} J_{12}(i,L)[\ 0 \quad p(i)I_m\ ]\Phi_V K_V(i,i)(\Phi_V^T)^{-i},$$

$$r_{3323}(L) = \sum_{i=1}^{L} J_{12}(i,L)[\ 0 \quad p(i)I_m\ ]R_U(i)(\Phi_V^T)^{-i},$$

$$r_{3331}(L)=\sum_{i=1}^{L}J_{13}(i,L)[\begin{array}{cc}0 & p(i)I_m\end{array}]\bar{H}K_{\bar{x}}(i,i)(\bar{\Phi})^{-i},$$

$$r_{3332}(L)=\sum_{i=1}^{L}J_{13}(i,L)[\begin{array}{cc}0 & p(i)I_m\end{array}]\Phi_V K_V(i,i)(\Phi_V^T)^{-i},$$

$$r_{3333}(L)=\sum_{i=1}^{L}J_{13}(i,L)[\begin{array}{cc}0 & p(i)I_m\end{array}]R_U(i)(\Phi_V^T)^{-i}.$$

Similarly, the equations for $J_2(L,L)$ and $J_3(L,L)$ are given by

$$\begin{aligned}J_2(L,L)=\{&\begin{bmatrix}\bar{\Phi}^{-L}K_{\bar{x}}(L,L)\bar{H}^T\\ \Phi_V^{-L}K_V(L,L)\Phi_V^T\\ \Phi_V^{-L}R_U(L)\end{bmatrix}\begin{bmatrix}0\\ I_m\end{bmatrix}\\ -r_{21}(L-1)&\begin{bmatrix}(\bar{\Phi}^T)\bar{H}^T\\ (\Phi_V^T)^{L+1}\\ (\Phi_V^T)^{L}\end{bmatrix}\begin{bmatrix}p_{2,2}(L)I_m\\ (1-p_{2,2}(L))I_m\end{bmatrix}\\ -r_{22}(L-1)&\begin{bmatrix}(\bar{\Phi}^T)\bar{H}^T\\ (\Phi_V^T)^{L+1}\\ (\Phi_V^T)^{L}\end{bmatrix}\begin{bmatrix}p(L)I_m\\ (1-p(L))I_m\end{bmatrix}\\ +r_{23}(L-1)&\begin{bmatrix}(\bar{\Phi}^T)\bar{H}^T\\ (\Phi_V^T)^{L+1}\\ (\Phi_V^T)^{L}\end{bmatrix}\begin{bmatrix}p_{2,2}(L)I_m\\ (1-p_{2,2}(L))I_m\end{bmatrix}\}\Gamma^{-1}(L),\end{aligned}\tag{A-13}$$

$$\begin{aligned}J_3(L,L)=\{&\begin{bmatrix}\bar{\Phi}^{-L}K_{\bar{x}}(L,L)\bar{H}^T\\ \Phi_V^{-L}K_V(L,L)\Phi_V^T\\ \Phi_V^{-L}R_U(L)\end{bmatrix}\begin{bmatrix}0\\ p(L)I_m\end{bmatrix}\\ -r_{31}(L-1)&\begin{bmatrix}(\bar{\Phi}^T)\bar{H}^T\\ (\Phi_V^T)^{L+1}\\ (\Phi_V^T)^{L}\end{bmatrix}\begin{bmatrix}p_{2,2}(L)I_m\\ (1-p_{2,2}(L))I_m\end{bmatrix}\\ -r_{32}(L-1)&\begin{bmatrix}(\bar{\Phi}^T)\bar{H}^T\\ (\Phi_V^T)^{L+1}\\ (\Phi_V^T)^{L}\end{bmatrix}\begin{bmatrix}p(L)I_m\\ (1-p(L))I_m\end{bmatrix}\\ +r_{33}(L-1)&\begin{bmatrix}(\bar{\Phi}^T)\bar{H}^T\\ (\Phi_V^T)^{L+1}\\ (\Phi_V^T)^{L}\end{bmatrix}\begin{bmatrix}p_{2,2}(L)I_m\\ (1-p_{2,2}(L))I_m\end{bmatrix}\}\Gamma^{-1}(L).\end{aligned}\tag{A-14}$$

From (A-7), it is shown that $r_{11}(L)$ satisfies

$$
\begin{aligned}
r_{11}(L) &= \sum_{i=1}^{L} J_1(i,L)[\ p(i)I_m \quad 0\ ][\ \bar{H}K_{\bar{x}}(i,i)(\bar{\Phi}^T)^{-i} \quad \Phi_V K_V(i,i)(\Phi_V^T)^{-i} \quad R_U(i)(\Phi_V^T)^{-i}\ ] \\
&= J_1(i,L)[\ p(L)I_m \quad 0\ ][\ \bar{H}K_{\bar{x}}(L,L)(\bar{\Phi}^T)^{-L} \quad \Phi_V K_V(L,L)(\Phi_V^T)^{-L} \quad R_U(L)(\Phi_V^T)^{-L}\ ] \\
&+ \sum_{i=1}^{L-1} J_1(i,L)[\ p(i)I_m \quad 0\ ][\ \bar{H}K_{\bar{x}}(i,i)(\bar{\Phi}^T)^{-i} \quad \Phi_V K_V(i,i)(\Phi_V^T)^{-i} \quad R_U(i)(\Phi_V^T)^{-i}\ ] \\
&= J_1(i,L)[\ p(L)I_m \quad 0\ ][\ \bar{H}K_{\bar{x}}(L,L)(\bar{\Phi}^T)^{-L} \quad \Phi_V K_V(L,L)(\Phi_V^T)^{-L} \quad R_U(L)(\Phi_V^T)^{-L}\ ] \\
&+ r_{11}(L-1) - J_1(L,L)\{[\ p_{2,2}(L)I_m \quad (1-p_{2,2}(L))I_m\ ][\ \bar{H}\bar{\Phi}^L \quad \Phi_V^{L+1} \quad \Phi_V^L\ ]r_{11}(L-1) \\
&+ [\ p(L)I_m \quad (1-p(L))I_m\ ][\ \bar{H}\bar{\Phi}^L \quad \Phi_V^{L+1} \quad \Phi_V^L\ ]r_{21}(L-1) \\
&- [\ p_{2,2}(L)I_m \quad (1-p_{2,2}(L))I_m\ ][\ \bar{H}\bar{\Phi}^L \quad \Phi_V^{L+1} \quad \Phi_V^L\ ]r_{31}(L-1)\}, \quad r_{11}(0) = 0.
\end{aligned}
\tag{A-15}
$$

Similarly, the equations for $r_{12}(L)$, $r_{13}(L)$, $r_{21}(L)$, $r_{22}(L)$, $r_{23}(L)$, $r_{31}(L)$, $r_{32}(L)$ and $r_{33}(L)$ are obtained as follows.

$$
\begin{aligned}
r_{12}(L) &= J_1(L,L)[\ 0 \quad I_m\ ][\ \bar{H}K_{\bar{x}}(L,L)(\bar{\Phi}^T)^{-L} \quad \Phi_V K_V(L,L)(\Phi_V^T)^{-L} \quad R_U(L)(\Phi_V^T)^{-L}\ ] \\
&+ r_{12}(L-1) - J_1(L,L)\{[\ p_{2,2}(L)I_m \quad (1-p_{2,2}(L))I_m\ ][\ \bar{H}\bar{\Phi}^L \quad \Phi_V^{L+1} \quad \Phi_V^L\ ]r_{12}(L-1) \\
&+ [\ p(L)I_m \quad (1-p(L))I_m\ ][\ \bar{H}\bar{\Phi}^L \quad \Phi_V^{L+1} \quad \Phi_V^L\ ]r_{22}(L-1) \\
&- [\ p_{2,2}(L)I_m \quad (1-p_{2,2}(L))I_m\ ][\ \bar{H}\bar{\Phi}^L \quad \Phi_V^{L+1} \quad \Phi_V^L\ ]r_{32}(L-1)\}, \quad r_{12}(0) = 0
\end{aligned}
$$

$$
\begin{aligned}
r_{13}(L) &= J_1(L,L)[\ 0 \quad p(L)I_m\ ][\ \bar{H}K_{\bar{x}}(L,L)(\bar{\Phi}^T)^{-L} \quad \Phi_V K_V(L,L)(\Phi_V^T)^{-L} \quad R_U(L)(\Phi_V^T)^{-L}\ ] \\
&+ r_{13}(L-1) - J_1(L,L)\{[\ p_{2,2}(L)I_m \quad (1-p_{2,2}(L))I_m\ ][\ \bar{H}\bar{\Phi}^L \quad \Phi_V^{L+1} \quad \Phi_V^L\ ]r_{13}(L-1) \\
&+ [\ p(L)I_m \quad (1-p(L))I_m\ ][\ \bar{H}\bar{\Phi}^L \quad \Phi_V^{L+1} \quad \Phi_V^L\ ]r_{23}(L-1) \\
&- [\ p_{2,2}(L)I_m \quad (1-p_{2,2}(L))I_m\ ][\ \bar{H}\bar{\Phi}^L \quad \Phi_V^{L+1} \quad \Phi_V^L\ ]r_{33}(L-1)\}, \quad r_{13}(0) = 0
\end{aligned}
$$

$$
\begin{aligned}
r_{21}(L) &= J_2(L,L)[\ p(L)I_m \quad 0\ ][\ \bar{H}K_{\bar{x}}(L,L)(\bar{\Phi}^T)^{-L} \quad \Phi_V K_V(L,L)(\Phi_V^T)^{-L} \quad R_U(L)(\Phi_V^T)^{-L}\ ] \\
&+ r_{21}(L-1) - J_2(L,L)\{[\ p_{2,2}(L)I_m \quad (1-p_{2,2}(L))I_m\ ][\ \bar{H}\bar{\Phi}^L \quad \Phi_V^{L+1} \quad \Phi_V^L\ ]r_{11}(L-1) \\
&+ [\ p(L)I_m \quad (1-p(L))I_m\ ][\ \bar{H}\bar{\Phi}^L \quad \Phi_V^{L+1} \quad \Phi_V^L\ ]r_{21}(L-1) \\
&- [\ p_{2,2}(L)I_m \quad (1-p_{2,2}(L))I_m\ ][\ \bar{H}\bar{\Phi}^L \quad \Phi_V^{L+1} \quad \Phi_V^L\ ]r_{31}(L-1)\}, \quad r_{21}(0) = 0
\end{aligned}
$$

$$
\begin{aligned}
r_{22}(L) &= J_2(L,L)[\ 0 \quad I_m\ ][\ \bar{H}K_{\bar{x}}(L,L)(\bar{\Phi}^T)^{-L} \quad \Phi_V K_V(L,L)(\Phi_V^T)^{-L} \quad R_U(L)(\Phi_V^T)^{-L}\ ] \\
&+ r_{22}(L-1) - J_2(L,L)\{[\ p_{2,2}(L)I_m \quad (1-p_{2,2}(L))I_m\ ][\ \bar{H}\bar{\Phi}^L \quad \Phi_V^{L+1} \quad \Phi_V^L\ ]r_{12}(L-1) \\
&+ [\ p(L)I_m \quad (1-p(L))I_m\ ][\ \bar{H}\bar{\Phi}^L \quad \Phi_V^{L+1} \quad \Phi_V^L\ ]r_{22}(L-1) \\
&- [\ p_{2,2}(L)I_m \quad (1-p_{2,2}(L))I_m\ ][\ \bar{H}\bar{\Phi}^L \quad \Phi_V^{L+1} \quad \Phi_V^L\ ]r_{32}(L-1)\}, \quad r_{22}(0) = 0
\end{aligned}
$$

$$
\begin{aligned}
r_{23}(L) &= J_2(L,L)[\ 0 \quad p(L)I_m\ ][\ \bar{H}K_{\bar{x}}(L,L)(\bar{\Phi}^T)^{-L} \quad \Phi_V K_V(L,L)(\Phi_V^T)^{-L} \quad R_U(L)(\Phi_V^T)^{-L}\ ] \\
&+ r_{23}(L-1) - J_2(L,L)\{[\ p_{2,2}(L)I_m \quad (1-p_{2,2}(L))I_m\ ][\ \bar{H}\bar{\Phi}^L \quad \Phi_V^{L+1} \quad \Phi_V^L\ ]r_{13}(L-1) \\
&+ [\ p(L)I_m \quad (1-p(L))I_m\ ][\ \bar{H}\bar{\Phi}^L \quad \Phi_V^{L+1} \quad \Phi_V^L\ ]r_{23}(L-1) \\
&- [\ p_{2,2}(L)I_m \quad (1-p_{2,2}(L))I_m\ ][\ \bar{H}\bar{\Phi}^L \quad \Phi_V^{L+1} \quad \Phi_V^L\ ]r_{33}(L-1)\}, \quad r_{23}(0) = 0
\end{aligned}
$$

$$r_{31}(L) = J_3(L,L)[\ p(L)I_m \quad 0\ ][\ \bar{H}K_{\tilde{x}}(L,L)(\bar{\Phi}^T)^{-L} \quad \Phi_V K_V(L,L)(\Phi_V^T)^{-L} \quad R_U(L)(\Phi_V^T)^{-L}\ ]$$
$$+r_{31}(L-1) - J_3(L,L)\{[\ p_{2,2}(L)I_m \quad (1-p_{2,2}(L))I_m\ ][\ \bar{H}\bar{\Phi}^L \quad \Phi_V^{L+1} \quad \Phi_V^L\ ]r_{11}(L-1)$$
$$+[\ p(L)I_m \quad (1-p(L))I_m\ ][\ \bar{H}\bar{\Phi}^L \quad \Phi_V^{L+1} \quad \Phi_V^L\ ]r_{21}(L-1)$$
$$-[\ p_{2,2}(L)I_m \quad (1-p_{2,2}(L))I_m\ ][\ \bar{H}\bar{\Phi}^L \quad \Phi_V^{L+1} \quad \Phi_V^L\ ]r_{31}(L-1)\}\ , \quad r_{31}(0) = 0$$

$$r_{32}(L) = J_3(L,L)[\ 0 \quad I_m\ ][\ \bar{H}K_{\tilde{x}}(L,L)(\bar{\Phi}^T)^{-L} \quad \Phi_V K_V(L,L)(\Phi_V^T)^{-L} \quad R_U(L)(\Phi_V^T)^{-L}\ ]$$
$$+r_{32}(L-1) - J_3(L,L)\{[\ p_{2,2}(L)I_m \quad (1-p_{2,2}(L))I_m\ ][\ \bar{H}\bar{\Phi}^L \quad \Phi_V^{L+1} \quad \Phi_V^L\ ]r_{12}(L-1)$$
$$+[\ p(L)I_m \quad (1-p(L))I_m\ ][\ \bar{H}\bar{\Phi}^L \quad \Phi_V^{L+1} \quad \Phi_V^L\ ]r_{22}(L-1)$$
$$-[\ p_{2,2}(L)I_m \quad (1-p_{2,2}(L))I_m\ ][\ \bar{H}\bar{\Phi}^L \quad \Phi_V^{L+1} \quad \Phi_V^L\ ]r_{32}(L-1)\}\ , \quad r_{32}(0) = 0$$

$$r_{33}(L) = J_3(L,L)[\ 0 \quad p(L)I_m\ ][\ \bar{H}K_{\tilde{x}}(L,L)(\bar{\Phi}^T)^{-L} \quad \Phi_V K_V(L,L)(\Phi_V^T)^{-L} \quad R_U(L)(\Phi_V^T)^{-L}\ ]$$
$$+r_{33}(L-1) - J_3(L,L)\{[\ p_{2,2}(L)I_m \quad (1-p_{2,2}(L))I_m\ ][\ \bar{H}\bar{\Phi}^L \quad \Phi_V^{L+1} \quad \Phi_V^L\ ]r_{13}(L-1)$$
$$+[\ p(L)I_m \quad (1-p(L))I_m\ ][\ \bar{H}\bar{\Phi}^L \quad \Phi_V^{L+1} \quad \Phi_V^L\ ]r_{23}(L-1)$$
$$-[\ p_{2,2}(L)I_m \quad (1-p_{2,2}(L))I_m\ ][\ \bar{H}\bar{\Phi}^L \quad \Phi_V^{L+1} \quad \Phi_V^L\ ]r_{33}(L-1)\}\ , \quad r_{33}(0) = 0$$

From (2.26), $h(k,L,L)$ is expressed as

$$h(k,L,L)\bar{R}_U(L) = K_{\tilde{x}}(k,L)\bar{H}^T\bar{p}^T(L) - h(k,L,L)E_\gamma[\bar{\gamma}(L)(\bar{H}K_{\tilde{x}}(L,L)\bar{H}^T$$
$$+\Phi_V K_V(L,L)\Phi_V^T + \Phi_V K_{VU}(L,L) + K_{UV}(L,L)\Phi_V^T)\bar{\gamma}^T(L)]$$
$$-\sum_{i=1}^{L-1} h(k,i,L)E_\gamma[\bar{\gamma}(i)(\bar{H}K_{\tilde{x}}(i,L)\bar{H}^T + \Phi_V K_V(i,L)\Phi_V^T$$
$$+\Phi_V K_{VU}(i,L) + K_{UV}(i,L)\Phi_V^T)\bar{\gamma}^T(L)].$$

Furthermore, the above equation is rewritten as follows.

$$h(k,L,L)\{\bar{R}_U(L) + [\ p(L)I_m \quad 0\ ][\bar{H}K_{\tilde{x}}(L,L)\bar{H}^T + \Phi_V K_V(L,L)\Phi_V^T]\begin{bmatrix} I_m \\ 0 \end{bmatrix}$$
$$+[\ 0 \quad (1-p(L))I_m\ ][\bar{H}K_{\tilde{x}}(L,L)\bar{H}^T + \Phi_V K_V(L,L)\Phi_V^T]\begin{bmatrix} 0 \\ I_m \end{bmatrix}\}$$
$$= K_{\tilde{x}}(k,L)\bar{H}^T\bar{p}^T(L)$$
$$-\sum_{i=1}^{L-1} h(k,i,L)E_\gamma\{\bar{\gamma}(i)(\bar{H}K_{\tilde{x}}(i,L)\bar{H}^T + \Phi_V K_V(i,i)(\Phi_V^T)^{-i}(\Phi_V^T)^{L+1} + R_U(i)(\Phi_V^T)^{-i}(\Phi_V^T)^L)\bar{\gamma}^T(L)\}$$

$$h(k,L,L)\Lambda(L) = K_{\bar{x}}(k,k)(\bar{\Phi}^T)^{L-k}\bar{H}^T \begin{bmatrix} p(L)I_m \\ (1-p(L))I_m \end{bmatrix}$$

$$-\sum_{i=1}^{L} h(k,i,L)\{[\ p(i)I_m \quad 0\ ][\ \bar{H}K_{\bar{x}}(i,i)(\bar{\Phi}^T)^{-i} \quad \Phi_V K_V(i,i)(\Phi_V^T)^{-i} \quad R_U(i)(\Phi_V^T)^{-i}\ ] \begin{bmatrix} (\bar{\Phi}^T)^L\bar{H}^T \\ (\Phi_V^T)^{L+1} \\ (\Phi_V^T)^L \end{bmatrix}$$

$$\begin{bmatrix} p_{2,2}(L)I_m \\ (1-p_{2,2}(L))I_m \end{bmatrix}$$

$$+[\ 0 \quad I_m\ ][\ \bar{H}K_{\bar{x}}(i,i)(\bar{\Phi}^T)^{-i} \quad \Phi_V K_V(i,i)(\Phi_V^T)^{-i} \quad R_U(i)(\Phi_V^T)^{-i}\ ] \begin{bmatrix} (\bar{\Phi}^T)^L\bar{H}^T \\ (\Phi_V^T)^{L+1} \\ (\Phi_V^T)^L \end{bmatrix}$$

$$\begin{bmatrix} p(L)I_m \\ (1-p(L))I_m \end{bmatrix}$$

$$-[\ 0 \quad p(i)I_m\ ][\ \bar{H}K_{\bar{x}}(i,i)(\bar{\Phi}^T)^{-i} \quad \Phi_V K_V(i,i)(\Phi_V^T)^{-i} \quad R_U(i)(\Phi_V^T)^{-i}\ ] \begin{bmatrix} (\bar{\Phi}^T)^L\bar{H}^T \\ (\Phi_V^T)^{L+1} \\ (\Phi_V^T)^L \end{bmatrix}$$

$$\begin{bmatrix} p_{2,2}(L)I_m \\ (1-p_{2,2}(L))I_m \end{bmatrix}\}$$

Let us introduce the functions $C_1(k,L)$, $C_2(k,L)$ and $C_3(k,L)$ as follows.

$$C_1(k,L) = \sum_{i=1}^{L} h(k,i,L)[\ p(i)I_m \quad 0\ ][\ \bar{H}K_{\bar{x}}(i,i)(\bar{\Phi}^T)^{-i} \quad \Phi_V K_V(i,i)(\Phi_V^T)^{-i} \quad R_U(i)(\Phi_V^T)^{-i}\ ]$$

$$C_2(k,L) = \sum_{i=1}^{L} h(k,i,L)[\ 0 \quad I_m\ ][\ \bar{H}K_{\bar{x}}(i,i)(\bar{\Phi}^T)^{-i} \quad \Phi_V K_V(i,i)(\Phi_V^T)^{-i} \quad R_U(i)(\Phi_V^T)^{-i}\ ]$$

$$C_3(k,L) = \sum_{i=1}^{L} h(k,i,L)\{[\ 0 \quad p(i)I_m\ ][\ \bar{H}K_{\bar{x}}(i,i)(\bar{\Phi}^T)^{-i} \quad \Phi_V K_V(i,i)(\Phi_V^T)^{-i} \quad R_U(i)(\Phi_V^T)^{-i}\ ]$$

If we subtract $C_1(k,L-1)$ from $C_1(k,L)$ and use (A-6), we obtain

$$\begin{aligned}
&C_1(k,L) - C_1(k,L-1) \\
&= h(k,L,L)[\ p(L)I_m \quad 0\ ][\ \bar{H}K_{\bar{x}}(L,L)\bar{\Phi}^{-L} \quad \Phi_V K_V(L,L)(\Phi_V^T)^{-L} \quad R_U(L)(\Phi_V^T)^{-L}\ ] \\
&- h(k,L,L)\{[\ p_{2,2}(L)I_m \quad (1-p_{2,2}(L))I_m\ ][\ \bar{H}\bar{\Phi}^L \quad \Phi_V^{L+1} \quad \Phi_V^L\ ]r_{11}(L-1) \\
&+ [\ p(L)I_m \quad (1-p(L))I_m\ ][\ \bar{H}\bar{\Phi}^L \quad \Phi_V^{L+1} \quad \Phi_V^L\ ]r_{21}(L-1) \\
&- [\ p_{2,2}(L)I_m \quad (1-p_{2,2}(L))I_m\ ][\ \bar{H}\bar{\Phi}^L \quad \Phi_V^{L+1} \quad \Phi_V^L\ ]r_{31}(L-1)\}.
\end{aligned}$$

Similarly, we obtain the following equations for $C_2(k,L)$ and $C_3(k,L)$ as

$$\begin{aligned}
&C_2(k,L) - C_2(k,L-1) \\
&= h(k,L,L)[\ 0 \quad I_m\ ][\ \bar{H}K_{\bar{x}}(L,L)\bar{\Phi}^{-L} \quad \Phi_V K_V(L,L)(\Phi_V^T)^{-L} \quad R_U(L)(\Phi_V^T)^{-L}\ ] \\
&- h(k,L,L)\{[\ p_{2,2}(L)I_m \quad (1-p_{2,2}(L))I_m\ ][\ \bar{H}\bar{\Phi}^L \quad \Phi_V^{L+1} \quad \Phi_V^L\ ]r_{12}(L-1) \\
&+ [\ p(L)I_m \quad (1-p(L))I_m\ ][\ \bar{H}\bar{\Phi}^L \quad \Phi_V^{L+1} \quad \Phi_V^L\ ]r_{22}(L-1) \\
&- [\ p_{2,2}(L)I_m \quad (1-p_{2,2}(L))I_m\ ][\ \bar{H}\bar{\Phi}^L \quad \Phi_V^{L+1} \quad \Phi_V^L\ ]r_{32}(L-1)\},
\end{aligned}$$

$$
\begin{aligned}
&C_3(k,L) - C_3(k,L-1) \\
&= h(k,L,L)[\ 0 \quad p(L)I_m\ ][\ \bar{H}K_{\bar{x}}(L,L)\bar{\Phi}^{-L} \quad \Phi_V K_V(L,L)(\Phi_V^T)^{-L} \quad R_U(L)(\Phi_V^T)^{-L}\ ] \\
&-h(k,L,L)\{[\ p_{2,2}(L)I_m \quad (1-p_{2,2}(L))I_m\ ][\ \bar{H}\bar{\Phi}^L \quad \Phi_V^{L+1} \quad \Phi_V^L\ ]r_{13}(L-1) \\
&+[\ p(L)I_m \quad (1-p(L))I_m\ ][\ \bar{H}\bar{\Phi}^L \quad \Phi_V^{L+1} \quad \Phi_V^L\ ]r_{23}(L-1) \\
&-[\ p_{2,2}(L)I_m \quad (1-p_{2,2}(L))I_m\ ][\ \bar{H}\bar{\Phi}^L \quad \Phi_V^{L+1} \quad \Phi_V^L\ ]r_{33}(L-1)\}.
\end{aligned}
$$

By denoting $C_1(k,L)$, $C_2(k,L)$ and $C_3(k,L)$ in terms of the matrix elements as

$$C_1(k,L) = [\ C_{11}(k,L) \quad C_{12}(k,L) \quad C_{13}(k,L)\ ],$$

$$C_2(k,L) = [\ C_{21}(k,L) \quad C_{22}(k,L) \quad C_{23}(k,L)\ ],$$

$$C_3(k,L) = [\ C_{31}(k,L) \quad C_{32}(k,L) \quad C_{33}(k,L)\ ],$$

$C_{ij}\ (k,L)$, $i = 1,2,3$, $j = 1,2,3$, are represented as follows.

$$C_{11}(k,L) = \sum_{i=1}^{L} h(k,i,L)[\ p(i)I_m \quad 0\ ]\bar{H}K_{\bar{x}}(i,i)(\bar{\Phi}^T)^{-i}$$

$$C_{12}(k,L) = \sum_{i=1}^{L} h(k,i,L)[\ p(i)I_m \quad 0\ ]\Phi_V K_V(i,i)(\Phi_V^T)^{-i}$$

$$C_{13}(k,L) = \sum_{i=1}^{L} h(k,i,L)[\ p(i)I_m \quad 0\ ]R_U(i)(\Phi_V^T)^{-i}$$

$$C_{21}(k,L) = \sum_{i=1}^{L} h(k,i,L)[\ 0 \quad I_m\ ]\bar{H}K_{\bar{x}}(i,i)(\bar{\Phi}^T)^{-i}$$

$$C_{22}(k,L) = \sum_{i=1}^{L} h(k,i,L)[\ 0 \quad I_m\ ]\Phi_V K_V(i,i)(\Phi_V^T)^{-i}$$

$$C_{23}(k,L) = \sum_{i=1}^{L} h(k,i,L)[\ 0 \quad I_m\ ]R_U(i)(\Phi_V^T)^{-i}$$

$$C_{31}(k,L) = \sum_{i=1}^{L} h(k,i,L)\{[\ 0 \quad p(i)I_m\ ]\bar{H}K_{\bar{x}}(i,i)(\bar{\Phi}^T)^{-i}$$

$$C_{32}(k,L) = \sum_{i=1}^{L} h(k,i,L)\{[\ 0 \quad p(i)I_m\ ]\Phi_V K_V(i,i)(\Phi_V^T)^{-i}$$

$$C_{33}(k,L) = \sum_{i=1}^{L} h(k,i,L)\{[\ 0 \quad p(i)I_m\ ]R_U(i)(\Phi_V^T)^{-i}$$

Hence, $h(k,L,L)$ is written as

$$h(k,L,L)\Lambda(L) = K_{\bar{x}}(k,k)(\bar{\Phi}^T)^{L-k}\bar{H}^T \begin{bmatrix} p(L)I_m \\ (1-p(L))I_m \end{bmatrix}$$
$$-\begin{bmatrix} C_{11}(k,L-1) & C_{12}(k,L-1) & C_{13}(k,L-1) \end{bmatrix} \begin{bmatrix} (\bar{\Phi}^T)^L\bar{H}^T \\ (\Phi_V^T)^{L+1} \\ (\Phi_V^T)^L \end{bmatrix} \begin{bmatrix} p_{2,2}(L)I_m \\ (1-p_{2,2}(L))I_m \end{bmatrix}$$
$$-\begin{bmatrix} C_{21}(k,L-1) & C_{22}(k,L-1) & C_{23}(k,L-1) \end{bmatrix} \begin{bmatrix} (\bar{\Phi}^T)^L\bar{H}^T \\ (\Phi_V^T)^{L+1} \\ (\Phi_V^T)^L \end{bmatrix} \begin{bmatrix} p(L)I_m \\ (1-p(L))I_m \end{bmatrix}$$
$$-\begin{bmatrix} C_{31}(k,L-1) & C_{32}(k,L-1) & C_{33}(k,L-1) \end{bmatrix} \begin{bmatrix} (\bar{\Phi}^T)^L\bar{H}^T \\ (\Phi_V^T)^{L+1} \\ (\Phi_V^T)^L \end{bmatrix} \begin{bmatrix} p_{2,2}(L)I_m \\ (1-p_{2,2}(L))I_m \end{bmatrix}\}.$$

Let us introduce the functions, $\bar{C}_{ij}(k,L), \quad i = 1,...,3, \quad j = 1,...,3,$ as

$$_{11}(k,L) = C_{11}(k,L)(\bar{\Phi}^T)^L, \quad \bar{C}_{12}(k,L) = C_{12}(k,L)(\Phi_V^T)^L, \quad \bar{C}_{13}(k,L) = C_{13}(k,L)(\Phi_V^T)^L,$$
$$\bar{C}_{21}(k,L) = C_{21}(k,L)(\bar{\Phi}^T)^L, \quad \bar{C}_{22}(k,L) = C_{22}(k,L)(\Phi_V^T)^L, \quad \bar{C}_{23}(k,L) = C_{23}(k,L)(\Phi_V^T)^L,$$
$$\bar{C}_{31}(k,L) = C_{31}(k,L)(\bar{\Phi}^T)^L, \quad \bar{C}_{32}(k,L) = C_{32}(k,L)(\Phi_V^T)^L, \quad \bar{C}_{33}(k,L) = C_{33}(k,L)(\Phi_V^T)^L.$$

Henceforth, the smoother gain is formulated as

$$h(k,L,L) = \{K_{\bar{x}}(k,k)(\bar{\Phi}^T)^{L-k}\bar{H}^T \begin{bmatrix} p(L)I_m \\ (1-p(L))I_m \end{bmatrix}$$
$$-[\bar{C}_{11}(k,L-1)\bar{\Phi}^T\bar{H}^T + \bar{C}_{12}(k,L)(\Phi_V^T)^2 + \bar{C}_{13}(k,L)\Phi_V^T] \begin{bmatrix} p_{2,2}(L)I_m \\ (1-p_{2,2}(L))I_m \end{bmatrix}$$
$$-[\bar{C}_{21}(k,L)\bar{\Phi}^T\bar{H}^T + \bar{C}_{22}(k,L)(\Phi_V^T)^2 + \bar{C}_{23}(k,L)\Phi_V^T] \begin{bmatrix} p(L)I_m \\ (1-p(L))I_m \end{bmatrix}$$
$$-[\bar{C}_{31}(k,L)\bar{\Phi}^T\bar{H}^T + \bar{C}_{32}(k,L)(\Phi_V^T)^2 + \bar{C}_{33}(k,L)\Phi_V^T] \begin{bmatrix} p_{2,2}(L)I_m \\ (1-p_{2,2}(L))I_m \end{bmatrix}\}\Gamma^{-1}(L).$$

Subtracting $\hat{\hat{x}}(k,L-1)$ from $\hat{\hat{x}}(k,L)$, we have

$$\hat{\hat{x}}(k,L)-\hat{\hat{x}}(k,L-1)=h(k,L,L)y(L)+\sum_{i=1}^{L-1}(h(k,i,L)-h(k,i,L-1))y(i)$$
$$=h(k,L,L)y(L)$$
$$-h(k,L,L)\{[\ p_{2,2}(L)I_m\quad (1-p_{2,2}(L))I_m\ ][\ \bar{H}\bar{\Phi}^L\quad \Phi_V^{L+1}\quad \Phi_V^L\ ]\sum_{i=1}^{L-1}J_1(i,L-1)$$
$$+[\ p(L)I_m\quad (1-p(L))I_m\ ][\ \bar{H}\bar{\Phi}^L\quad \Phi_V^{L+1}\quad \Phi_V^L\ ]\sum_{i=1}^{L-1}J_2(i,L-1)$$
$$-[\ p_{2,2}(L)I_m\quad (1-p_{2,2}(L))I_m\ ][\ \bar{H}\bar{\Phi}^L\quad \Phi_V^{L+1}\quad \Phi_V^L\ ]\sum_{i=1}^{L-1}J_3(i,L-1)\}y(i).$$

By introducing the functions

$$O_1(L)=\sum_{i=1}^{L}J_1(i,L)y(i),\quad O_2(L)=\sum_{i=1}^{L}J_2(i,L)y(i),\quad O_3(L)=\sum_{i=1}^{L}J_3(i,L)y(i),$$

we obtain the equation updating the fixed-point smoothing estimate $\hat{\hat{x}}(k,L)$ as

$$\begin{aligned}&\hat{\hat{x}}(k,L)-\hat{\hat{x}}(k,L-1)=h(k,L,L)\{y(L)\\&-[\ p_{2,2}(L)I_m\quad (1-p_{2,2}(L))I_m\ ][\ \bar{H}\bar{\Phi}^L\quad \Phi_V^{L+1}\quad \Phi_V^L\ ]O_1(L-1)\\&-[\ p(L)I_m\quad (1-p(L))I_m\ ][\ \bar{H}\bar{\Phi}^L\quad \Phi_V^{L+1}\quad \Phi_V^L\ ]O_2(L-1)\\&+[\ p_{2,2}(L)I_m\quad (1-p_{2,2}(L))I_m\ ][\ \bar{H}\bar{\Phi}^L\quad \Phi_V^{L+1}\quad \Phi_V^L\ ]O_3(L-1)\}.\end{aligned}\tag{A-16}$$

Subtracting $O_1(L-1)$ from $O_1(L)$ and using (A-7), we obtain

$$\begin{aligned}&O_1(L)-O_1(L-1)=J_1(L,L)y(L)+\sum_{i=1}^{L-1}(J_1(i,L)-J_1(i,L-1))\\&=J_1(L,L)\{y(L)-[\ p_{2,2}(L)I_m\quad (1-p_{2,2}(L))I_m\ ][\ \bar{H}\bar{\Phi}^L\quad \Phi_V^{L+1}\quad \Phi_V^L\ ]O_1(L-1)\\&-[\ p(L)I_m\quad (1-p(L))I_m\ ][\ \bar{H}\bar{\Phi}^L\quad \Phi_V^{L+1}\quad \Phi_V^L\ ]O_2(L-1)\\&+[\ p_{2,2}(L)I_m\quad (1-p_{2,2}(L))I_m\ ][\ \bar{H}\bar{\Phi}^L\quad \Phi_V^{L+1}\quad \Phi_V^L\ ]O_3(L-1)\}.\end{aligned}\tag{A-17}$$

Similarly, we have the equations updating $O_2(L)$ and $O_3(L)$ as

$$\begin{aligned}&O_2(L)-O_2(L-1)=J_2(L,L)\{y(L)\\&-[\ p_{2,2}(L)I_m\quad (1-p_{2,2}(L))I_m\ ][\ \bar{H}\bar{\Phi}^L\quad \Phi_V^{L+1}\quad \Phi_V^L\ ]O_1(L-1)\\&-[\ p(L)I_m\quad (1-p(L))I_m\ ][\ \bar{H}\bar{\Phi}^L\quad \Phi_V^{L+1}\quad \Phi_V^L\ ]O_2(L-1)\\&+[\ p_{2,2}(L)I_m\quad (1-p_{2,2}(L))I_m\ ][\ \bar{H}\bar{\Phi}^L\quad \Phi_V^{L+1}\quad \Phi_V^L\ ]O_3(L-1)\},\end{aligned} \tag{A-18}$$

$$\begin{aligned}&O_3(L)-O_3(L-1)=J_3(L,L)\{y(L)\\&-[\ p_{2,2}(L)I_m\quad (1-p_{2,2}(L))I_m\ ][\ \bar{H}\bar{\Phi}^L\quad \Phi_V^{L+1}\quad \Phi_V^L\ ]O_1(L-1)\\&-[\ p(L)I_m\quad (1-p(L))I_m\ ][\ \bar{H}\bar{\Phi}^L\quad \Phi_V^{L+1}\quad \Phi_V^L\ ]O_2(L-1)\\&+[\ p_{2,2}(L)I_m\quad (1-p_{2,2}(L))I_m\ ][\ \bar{H}\bar{\Phi}^L\quad \Phi_V^{L+1}\quad \Phi_V^L\ ]O_3(L-1)\}.\end{aligned} \tag{A-19}$$

From (2.17), the filtering estimate of $\bar{x}(k)$ is given by

$$\hat{\bar{x}}(L,L)=\sum_{i=1}^{L}h(L,i,L)y(i).$$

From (2.26), $h(L,s,L)$ satisfies

$$\begin{aligned}&h(L,s,L)\bar{R}_U(s)=K_{\bar{x}}(L,s)\bar{H}^T\bar{p}^T(s)\\&-\sum_{i=1}^{L}h(L,i,L)E_\gamma[\bar{\gamma}(i)(\bar{H}K_{\bar{x}}(i,s)\bar{H}^T+\Phi_V K_V(i,s)\Phi_V^T+\Phi_V K_{VU}(i,s)+K_{UV}(i,s)\Phi_V^T)\bar{\gamma}^T(s)].\end{aligned}$$

Let us introduce the equation for $\bar{J}(s,L)$, which satisfies

$$\begin{aligned}&\bar{J}(s,L)\bar{R}_U(s)=\bar{\Phi}^{-s}K_{\bar{x}}(s,s)\bar{H}^T\bar{p}^T(s)\\&-\sum_{i=1}^{L}\bar{J}(i,L)E_\gamma[\bar{\gamma}(i)(\bar{H}K_{\bar{x}}(i,s)\bar{H}^T+\Phi_V K_V(i,s)\Phi_V^T\\&+\Phi_V K_{VU}(i,s)+K_{UV}(i,s)\Phi_V^T)\bar{\gamma}^T(s)].\end{aligned} \tag{A-20}$$

Hence, $h(L,s,L)$ is formulated as

$$h(L,s,L)=\bar{\Phi}^L\bar{J}(s,L). \tag{A-21}$$

Introducing the function

$$e(L)=\sum_{i=1}^{L}\bar{J}(i,L)y(i), \tag{A-22}$$

the filtering estimate $\hat{\hat{x}}(L,L)$ is expressed as

$$\hat{\hat{x}}(L,L) = \bar{\Phi}^L e(L). \tag{A-23}$$

By the way, from $K_{UV}(L,s) = 0$, the following relationship is derived.

$$\begin{aligned}
&E_\gamma[\bar{\gamma}(L)(\bar{H}K_{\bar{x}}(L,s)\bar{H}^T + \Phi_V K_V(L,s)\Phi_V^T + \Phi_V K_{VU}(L,s) + K_{UV}(L,s)\Phi_V^T)\bar{\gamma}^T(s)]\\
&= [\ p_{2,2}(L)I_m \quad (1-p_{2,2}(L))I_m\ ][\bar{H}K_{\bar{x}}(L,s)\bar{H}^T + \Phi_V K_V(L,s)\Phi_V^T\\
&+\Phi_V K_{VU}(L,s) + K_{UV}(L,s)\Phi_V^T]\begin{bmatrix} p(s)I_m \\ 0 \end{bmatrix}\\
&+[\ p(L)I_m \quad (1-p(L))I_m\ ][\bar{H}K_{\bar{x}}(L,s)\bar{H}^T + \Phi_V K_V(L,s)\Phi_V^T\\
&+\Phi_V K_{VU}(L,s) + K_{UV}(L,s)\Phi_V^T]\begin{bmatrix} 0 \\ I_m \end{bmatrix}\\
&-[\ p_{2,2}(L)I_m \quad (1-p_{2,2}(L))I_m\ ][\bar{H}K_{\bar{x}}(L,s)\bar{H}^T + \Phi_V K_V(L,s)\Phi_V^T\\
&+\Phi_V K_{VU}(L,s) + K_{UV}(L,s)\Phi_V^T]\begin{bmatrix} 0 \\ p(s)I_m \end{bmatrix}\\
&= [\ p_{2,2}(L)I_m \quad (1-p_{2,2}(L))I_m\ ][\bar{H}K_{\bar{x}}(L,s)\bar{H}^T + \Phi_V K_V(L,s)\Phi_V^T + \Phi_V^{L-s}R_U(s)]\begin{bmatrix} p(s)I_m \\ 0 \end{bmatrix}\\
&+[\ p(L)I_m \quad (1-p(L))I_m\ ][\bar{H}K_{\bar{x}}(L,s)\bar{H}^T + \Phi_V K_V(L,s)\Phi_V^T + \Phi_V^{L-s}R_U(s)]\begin{bmatrix} 0 \\ I_m \end{bmatrix}\\
&-[\ p_{2,2}(L)I_m \quad (1-p_{2,2}(L))I_m\ ][\bar{H}K_{\bar{x}}(L,s)\bar{H}^T + \Phi_V K_V(L,s)\Phi_V^T + \Phi_V^{L-s}R_U(s)]\begin{bmatrix} 0 \\ p(s)I_m \end{bmatrix}
\end{aligned} \tag{5.1}$$

(A-24)

Subtracting the equation obtained by putting $L \to L-1$ in (A-20) from (A-20), we have

$$\begin{aligned}
&(\bar{J}(s,L) - \bar{J}(s,L-1))\bar{R}_U(s) = -\bar{J}(L,L)E_\gamma[\bar{\gamma}(L)(\bar{H}K_{\bar{x}}(L,s)\bar{H}^T\\
&+\Phi_V K_V(L,s)\Phi_V^T + \Phi_V K_{VU}(L,s) + K_{UV}(L,s)\Phi_V^T)\bar{\gamma}^T(s)]\\
&-\sum_{i=1}^{L-1} \bar{J}(i,L)E_\gamma[\bar{\gamma}(i)(\bar{H}K_{\bar{x}}(i,s)\bar{H}^T + \Phi_V K_V(i,s)\Phi_V^T + \Phi_V K_{VU}(i,s) + K_{UV}(i,s)\Phi_V^T)\bar{\gamma}^T(s)].
\end{aligned}$$

From (A-2), (A-3) and (A-4), it is shown that $\bar{J}(s,L)$ is updated by

$$\begin{aligned}
&\bar{J}(s,L) - \bar{J}(s,L-1) =\\
&-\bar{J}(L,L)\{[\ p_{2,2}(L)I_m \quad (1-p_{2,2}(L))I_m\ ][\ \bar{H}\bar{\Phi}^L \quad \Phi_V^{L+1} \quad \Phi_V^L\ ]J_1(s,L-1)\\
&+[\ p(L)I_m \quad (1-p(L))I_m 0\ ][\ \bar{H}\bar{\Phi}^L \quad \Phi_V^{L+1} \quad \Phi_V^L\ ]J_2(s,L-1)\\
&-[\ p_{2,2}(L)I_m \quad (1-p_{2,2}(L))I_m\ ][\ \bar{H}\bar{\Phi}^L \quad \Phi_V^{L+1} \quad \Phi_V^L\ ]J_3(s,L-1)\}.
\end{aligned} \tag{A-25}$$

Subtracting the equation obtained by putting $L \to L-1$ in (A-22) from (A-22), we obtain

$$e(L)-e(L-1)=\bar{J}(L,L)y(L)+\sum_{i=1}^{L-1}(\bar{J}(i,L)-\bar{J}(i,L-1))y(i)$$
$$=\bar{J}(L,L)y(L)-\bar{J}(L,L)\{[\ p_{2,2}(L)I_m \quad (1-p_{2,2}(L))I_m\ ][\ \bar{H}\bar{\Phi}^L \quad \Phi_V^{L+1} \quad \Phi_V^L\ ]O_1(s,L-1)$$
$$+[\ p(L)I_m \quad (1-p(L))I_m 0\ ][\ \bar{H}\bar{\Phi}^L \quad \Phi_V^{L+1} \quad \Phi_V^L\ ]O_2(s,L-1)$$
$$-[\ p_{2,2}(L)I_m \quad (1-p_{2,2}(L))I_m\ ][\ \bar{H}\bar{\Phi}^L \quad \Phi_V^{L+1} \quad \Phi_V^L\ ]O_3(s,L-1)\}. \tag{A-26}$$

From (A-20), $\bar{J}(L,L)$ satisfies the following equation.

$$\bar{J}(L,L)\bar{R}_U(L)=\bar{\Phi}^{-L}K_{\bar{x}}(L,L)\bar{H}^T\bar{p}^T(L)$$
$$-\sum_{i=1}^{L}\bar{J}(i,L)E_\gamma[\bar{\gamma}(i)(\bar{H}K_{\bar{x}}(i,L)\bar{H}^T+\Phi_V K_V(i,L)\Phi_V^T+\Phi_V K_{VU}(i,L)+K_{UV}(i,L)\Phi_V^T)\bar{\gamma}^T(L)]$$
$$=\bar{\Phi}^{-L}K_{\bar{x}}(L,L)\bar{H}^T\bar{p}^T(L)-\bar{J}(L,L)E_\gamma[\bar{\gamma}(L)(\bar{H}K_{\bar{x}}(L,L)\bar{H}^T+\Phi_V K_V(L,L)\Phi_V^T$$
$$+\Phi_V K_{VU}(L,L)+K_{UV}(L,L)\Phi_V^T)\bar{\gamma}^T(L)]$$
$$-\sum_{i=1}^{L-1}\bar{J}(i,L)E_\gamma[\bar{\gamma}(i)(\bar{H}K_{\bar{x}}(i,L)\bar{H}^T+\Phi_V K_V(i,L)\Phi_V^T+\Phi_V K_{VU}(i,L)+K_{UV}(i,L)\Phi_V^T)\bar{\gamma}^T(L)] \tag{5.2}$$

$$=\bar{\Phi}^{-L}K_{\bar{x}}(L,L)\bar{H}^T\bar{p}^T(L)-\bar{J}(L,L)\{[\ p(L)I_m \quad 0\ ](\bar{H}K_{\bar{x}}(L,L)\bar{H}^T+\Phi_V K_V(L,L)\Phi_V^T)\begin{bmatrix} I_m \\ 0 \end{bmatrix}$$
$$[\ 0 \quad (1-p(L))I_m\ ]\bar{H}K_{\bar{x}}(L,L)\bar{H}^T+\Phi_V K_V(L,L)\Phi_V^T)\begin{bmatrix} 0 \\ I_m \end{bmatrix}\}$$
$$-\sum_{i=1}^{L}\{\bar{J}(i,L-1)-\bar{J}(L,L)\{[\ p_{2,2}(L)I_m \quad (1-p_{2,2}(L))I_m\ ][\ \bar{H}\bar{\Phi}^L \quad \Phi_V^{L+1} \quad \Phi_V^L\ ]J_1(i,L-1)$$
$$+[\ p(L)I_m \quad (1-p(L))I_m 0\ ][\ \bar{H}\bar{\Phi}^L \quad \Phi_V^{L+1} \quad \Phi_V^L\ ]J_2(i,L-1)$$
$$-[\ p_{2,2}(L)I_m \quad (1-p_{2,2}(L))I_m\ ][\ \bar{H}\bar{\Phi}^L \quad \Phi_V^{L+1} \quad \Phi_V^L\ ]J_3(i,L-1)\}$$
$$\{[\ p(i)I_m \quad 0\ ][\ \bar{H}K_{\bar{x}}(i,i)(\bar{\Phi}^T)^{-i} \quad \Phi_V K_V(i,i)(\Phi_V^T)^{-i} \quad R_U(i)(\Phi_V^T)^{-i}\ ]$$
$$\begin{bmatrix} (\bar{\Phi}^T)^L\bar{H}^T \\ (\Phi_V^T)^{L+1} \\ (\Phi_V^T)^L \end{bmatrix}\begin{bmatrix} p_{2,2}(L)I_m \\ (1-p_{2,2}(L))I_m \end{bmatrix}$$
$$+[\ 0 \quad I_m\ ][\ \bar{H}K_{\bar{x}}(i,i)(\bar{\Phi}^T)^{-i} \quad \Phi_V K_V(i,i)(\Phi_V^T)^{-i} \quad R_U(i)(\Phi_V^T)^{-i}\ ]$$
$$\begin{bmatrix} (\bar{\Phi}^T)^L\bar{H}^T \\ (\Phi_V^T)^{L+1} \\ (\Phi_V^T)^L \end{bmatrix}\begin{bmatrix} p(L)I_m \\ (1-p(L))I_m \end{bmatrix}$$
$$-[\ 0 \quad p(i)I_m\ ][\ \bar{H}K_{\bar{x}}(i,i)(\bar{\Phi}^T)^{-i} \quad \Phi_V K_V(i,i)(\Phi_V^T)^{-i} \quad R_U(i)(\Phi_V^T)^{-i}\ ]$$
$$\begin{bmatrix} (\bar{\Phi}^T)^L\bar{H}^T \\ (\Phi_V^T)^{L+1} \\ (\Phi_V^T)^L \end{bmatrix}\begin{bmatrix} p_{2,2}(L)I_m \\ (1-p_{2,2}(L))I_m \end{bmatrix}\} \tag{A-27}$$

Introducing the functions $B_1(L)$, $B_2(L)$ and $B_2(L)$ as

$$B_1(L)=\sum_{i=1}^{L}\bar{J}(i,L)[\ p(i)I_m \quad 0\ ][\ \bar{H}K_{\bar{x}}(i,i)(\bar{\Phi}^T)^{-i} \quad \Phi_V K_V(i,i)(\Phi_V^T)^{-i} \quad R_U(i)(\Phi_V^T)^{-i}\ ],$$

$$B_2(L)=\sum_{i=1}^{L}\bar{J}(i,L)\begin{bmatrix} 0 & I_m \end{bmatrix}\begin{bmatrix} \bar{H}K_{\bar{x}}(i,i)(\bar{\Phi}^T)^{-i} & \Phi_V K_V(i,i)(\Phi_V^T)^{-i} & R_U(i)(\Phi_V^T)^{-i} \end{bmatrix},$$

$$B_3(L)=\sum_{i=1}^{L}\bar{J}(i,L)\begin{bmatrix} 0 & p(i)I_m \end{bmatrix}\begin{bmatrix} \bar{H}K_{\bar{x}}(i,i)(\bar{\Phi}^T)^{-i} & \Phi_V K_V(i,i)(\Phi_V^T)^{-i} & R_U(i)(\Phi_V^T)^{-i} \end{bmatrix},$$

after some manipulations, we obtain

$$\bar{J}(L,L)\Gamma(L)=\bar{\Phi}^{-L}K_{\bar{x}}(L,L)\bar{H}\bar{p}^T(L)-B_1(L-1)\begin{bmatrix} (\bar{\Phi}^T)^L\bar{H}^T \\ (\Phi_V^T)^{L+1} \\ (\Phi_V^T)^L \end{bmatrix}\begin{bmatrix} p_{2,2}(L)I_m \\ (1-p_{2,2}(L))I_m \end{bmatrix}$$
$$-B_2(L-1)\begin{bmatrix} (\bar{\Phi}^T)^L\bar{H}^T \\ (\Phi_V^T)^{L+1} \\ (\Phi_V^T)^L \end{bmatrix}\begin{bmatrix} p(L)I_m \\ (1-p(L))I_m \end{bmatrix}+B_3(L-1)\begin{bmatrix} (\bar{\Phi}^T)^L\bar{H}^T \\ (\Phi_V^T)^{L+1} \\ (\Phi_V^T)^L \end{bmatrix}\begin{bmatrix} p_{2,2}(L)I_m \\ (1-p_{2,2}(L))I_m \end{bmatrix}.$$

By putting

$$O_1(L)=\begin{bmatrix} O_{11}(L) \\ O_{12}(L) \\ O_{13}(L) \end{bmatrix},\quad O_2(L)=\begin{bmatrix} O_{21}(L) \\ O_{22}(L) \\ O_{23}(L) \end{bmatrix},\quad O_3(L)=\begin{bmatrix} O_{31}(L) \\ O_{32}(L) \\ O_{33}(L) \end{bmatrix},$$

$$\bar{O}_{11}(L)=\bar{\Phi}^L O_{11}(L),\quad \bar{O}_{12}(L)=\Phi_V^L O_{12}(L),\quad \bar{O}_{13}(L)=\Phi_V^L O_{13}(L),$$

$$\bar{O}_{21}(L)=\bar{\Phi}^L O_{21}(L),\quad \bar{O}_{22}(L)=\Phi_V^L O_{22}(L),\quad \bar{O}_{23}(L)=\Phi_V^L O_{23}(L),$$

$$\bar{O}_{31}(L)=\bar{\Phi}^L O_{31}(L),\quad \bar{O}_{32}(L)=\Phi_V^L O_{32}(L),\quad \bar{O}_{33}(L)=\Phi_V^L O_{23}(L),$$

the filtering estimate $\hat{\bar{x}}(L,L)$ is updated by

$$\begin{aligned}
&\hat{\bar{x}}(L,L)=\Phi\hat{\bar{x}}(L-1,L-1)+\bar{G}(L,L)\{y(L)\\
&-\begin{bmatrix} p_{2,2}(L)I_m & (1-p_{2,2}(L))I_m \end{bmatrix}\begin{bmatrix} \bar{H}\bar{\Phi}^L & \Phi_V^{L+1} & \Phi_V^L \end{bmatrix}O_1(s,L-1)\\
&-\begin{bmatrix} p(L)I_m & (1-p(L))I_m 0 \end{bmatrix}\begin{bmatrix} \bar{H}\bar{\Phi}^L & \Phi_V^{L+1} & \Phi_V^L \end{bmatrix}O_2(s,L-1)\\
&+\begin{bmatrix} p_{2,2}(L)I_m & (1-p_{2,2}(L))I_m \end{bmatrix}\begin{bmatrix} \bar{H}\bar{\Phi}^L & \Phi_V^{L+1} & \Phi_V^L \end{bmatrix}O_3(s,L-1)\}\\
&=\Phi\hat{x}(L-1,L-1)+\bar{G}(L,L)\{y(L)\\
&-\begin{bmatrix} p_{2,2}(L)I_m & (1-p_{2,2}(L))I_m \end{bmatrix}[\bar{H}\bar{\Phi}^L\bar{O}_{11}(L-1)+\Phi_V^2\bar{O}_{12}(L-1)+\Phi_V\bar{O}_{13}(L-1)]\\
&-\begin{bmatrix} p(L)I_m & (1-p(L))I_m 0 \end{bmatrix}[\bar{H}\bar{\Phi}^L\bar{O}_{21}(L-1)+\Phi_V^2\bar{O}_{22}(L-1)+\Phi_V\bar{O}_{23}(L-1)]\\
&+\begin{bmatrix} p_{2,2}(L)I_m & (1-p_{2,2}(L))I_m \end{bmatrix}[\bar{H}\bar{\Phi}^L\bar{O}_{31}(L-1)+\Phi_V^2\bar{O}_{32}(L-1)+\Phi_V\bar{O}_{33}(L-1)]\},\\
&\quad\hat{\bar{x}}(0,0)=0\ .
\end{aligned}\tag{A-28}$$

Also, by putting

$$B_1(L) = [\ B_{11}(L) \quad B_{12}(L) \quad B_{13}(L)\ ],$$

$$B_2(L) = [\ B_{21}(L) \quad B_{22}(L) \quad B_{23}(L)\ ],$$

$$B_3(L) = [\ B_{31}(L) \quad B_{32}(L) \quad B_{33}(L)\ ],$$

$$B_{11}(L) = \sum_{i=1}^{L} \bar{J}(i,L)[\ p(i)I_m \quad 0\ ]\bar{H}K_{\bar{x}}(i,i)(\bar{\Phi}^T)^{-i},$$

$$B_{12}(L) = \sum_{i=1}^{L} \bar{J}(i,L)[\ p(i)I_m \quad 0\ ]\Phi_V K_V(i,i)(\Phi_V^T)^{-i},$$

$$B_{13}(L) = \sum_{i=1}^{L} \bar{J}(i,L)[\ p(i)I_m \quad 0\ ]R_U(i)(\Phi_V^T)^{-i},$$

$$B_{21}(L) = \sum_{i=1}^{L} \bar{J}(i,L)[\ 0 \quad I_m\ ]\bar{H}K_{\bar{x}}(i,i)(\bar{\Phi}^T)^{-i},$$

$$B_{22}(L) = \sum_{i=1}^{L} \bar{J}(i,L)[\ 0 \quad I_m\ ]\Phi_V K_V(i,i)(\Phi_V^T)^{-i},$$

$$B_{23}(L) = \sum_{i=1}^{L} \bar{J}(i,L)[\ 0 \quad I_m\ ]R_U(i)(\Phi_V^T)^{-i},$$

$$B_{31}(L) = \sum_{i=1}^{L} \bar{J}(i,L)[\ 0 \quad p(i)I_m\ ]\bar{H}K_{\bar{x}}(i,i)(\bar{\Phi}^T)^{-i},$$

$$B_{32}(L) = \sum_{i=1}^{L} \bar{J}(i,L)[\ 0 \quad p(i)I_m\ ]\Phi_V K_V(i,i)(\Phi_V^T)^{-i},$$

$$B_{33}(L) = \sum_{i=1}^{L} \bar{J}(i,L)[\ 0 \quad p(i)I_m\ ]R_U(i)(\Phi_V^T)^{-i},$$

the filter gain $\bar{G}(L,L)$ is represented by

$$
\begin{aligned}
\bar{G}(L,L) &= \{K_{\bar{x}}(L,L)\bar{H}\bar{p}^T(L) - \bar{\Phi}^L B_1(L-1)\begin{bmatrix} (\bar{\Phi}^T)^L\bar{H}^T \\ (\Phi_V^T)^{L+1} \\ (\Phi_V^T)^L \end{bmatrix}\begin{bmatrix} p_{2,2}(L)I_m \\ (1-p_{2,2}(L))I_m \end{bmatrix} \\
&-\bar{\Phi}^L B_2(L-1)\begin{bmatrix} (\bar{\Phi}^T)^L\bar{H}^T \\ (\Phi_V^T)^{L+1} \\ (\Phi_V^T)^L \end{bmatrix}\begin{bmatrix} p(L)I_m \\ (1-p(L))I_m \end{bmatrix} \\
&+\bar{\Phi}^L B_3(L-1)\begin{bmatrix} (\bar{\Phi}^T)^L\bar{H}^T \\ (\Phi_V^T)^{L+1} \\ (\Phi_V^T)^L \end{bmatrix}\begin{bmatrix} p_{2,2}(L)I_m \\ (1-p_{2,2}(L))I_m \end{bmatrix}\}\Gamma^{-1}(L) \\
&= \{K_{\bar{x}}(L,L)\bar{H}\bar{p}^T(L) - [\bar{\Phi}^L B_{11}(L-1)(\bar{\Phi}^T)^L\bar{H}^T + \bar{\Phi}^L B_{12}(L-1)(\Phi_V^T)^{L+1} \\
&+\bar{\Phi}^L B_{13}(L-1)(\Phi_V^T)^L]\begin{bmatrix} p_{2,2}(L)I_m \\ (1-p_{2,2}(L))I_m \end{bmatrix} \\
&-[\bar{\Phi}^L B_{21}(L-1)(\bar{\Phi}^T)^L\bar{H}^T + \bar{\Phi}^L B_{22}(L-1)(\Phi_V^T)^{L+1} + \bar{\Phi}^L B_{23}(L-1)(\Phi_V^T)^L]\begin{bmatrix} p(L)I_m \\ (1-p(L))I_m \end{bmatrix} \\
&+[\bar{\Phi}^L B_{31}(L-1)(\bar{\Phi}^T)^L\bar{H}^T + \bar{\Phi}^L B_{32}(L-1)(\Phi_V^T)^{L+1} \\
&+\bar{\Phi}^L B_{33}(L-1)(\Phi_V^T)^L]\begin{bmatrix} p_{2,2}(L)I_m \\ (1-p_{2,2}(L))I_m \end{bmatrix}\}\Gamma^{-1}(L) \\
&= \{K_{\bar{x}}(L,L)\bar{H}\bar{p}^T(L) - [\bar{\Phi}\tilde{B}_{11}(L-1)\bar{\Phi}^T\bar{H}^T + \bar{\Phi}\tilde{B}_{12}(L-1)(\Phi_V^T)^2 \\
&+\bar{\Phi}\tilde{B}_{13}(L-1)\Phi_V^T]\begin{bmatrix} p_{2,2}(L)I_m \\ (1-p_{2,2}(L))I_m \end{bmatrix} \\
&-[\bar{\Phi}\tilde{B}_{21}(L-1)\bar{\Phi}^T\bar{H}^T + \bar{\Phi}\tilde{B}_{22}(L-1)(\Phi_V^T)^2 + \bar{\Phi}\tilde{B}_{23}(L-1)\Phi_V^T]\begin{bmatrix} p(L)I_m \\ (1-p(L))I_m \end{bmatrix} \\
&+[\bar{\Phi}\tilde{B}_{31}(L-1)\bar{\Phi}^T\bar{H}^T + \bar{\Phi}\tilde{B}_{32}(L-1)(\Phi_V^T)^2 + \bar{\Phi}\tilde{B}_{33}(L-1)\Phi_V^T]\begin{bmatrix} p_{2,2}(L)I_m \\ (1-p_{2,2}(L))I_m \end{bmatrix}\}\Gamma^{-1}(L).
\end{aligned}
\tag{A-29}
$$

Here, the following relationships for $\tilde{B}_{11}(L), \tilde{B}_{12}(L), \tilde{B}_{13}(L), \tilde{B}_{21}(L), \tilde{B}_{22}(L), \tilde{B}_{23}(L), \tilde{B}_{31}(L)$, $\tilde{B}_{32}(L)$ and $\tilde{B}_{33}(L)$ are introduced.

$$\tilde{B}_{11}(L) = \bar{\Phi}^L B_{11}(L)(\bar{\Phi}^T)^L, \quad \tilde{B}_{12}(L) = \bar{\Phi}^L B_{12}(L)(\Phi_V^T)^L, \quad \tilde{B}_{13}(L) = \bar{\Phi}^L B_{13}(L)(\Phi_V^T)^L,$$

$$\tilde{B}_{21}(L) = \bar{\Phi}^L B_{21}(L)(\bar{\Phi}^T)^L, \quad \tilde{B}_{22}(L) = \bar{\Phi}^L B_{22}(L)(\Phi_V^T)^L, \quad \tilde{B}_{23}(L) = \bar{\Phi}^L B_{23}(L)(\Phi_V^T)^L,$$

$$\tilde{B}_{31}(L) = \bar{\Phi}^L B_{31}(L)(\bar{\Phi}^T)^L, \quad \tilde{B}_{32}(L) = \bar{\Phi}^L B_{32}(L)(\Phi_V^T)^L, \quad \tilde{B}_{33}(L) = \bar{\Phi}^L B_{33}(L)(\Phi_V^T)^L.$$

By putting

$$J_1(s,L)=\begin{bmatrix} J_{11}(s,L)\\ J_{12}(s,L)\\ J_{13}(s,L)\end{bmatrix},\quad J_2(s,L)=\begin{bmatrix} J_{21}(s,L)\\ J_{22}(s,L)\\ J_{23}(s,L)\end{bmatrix},\quad J_3(s,L)=\begin{bmatrix} J_{31}(s,L)\\ J_{32}(s,L)\\ J_{33}(s,L)\end{bmatrix},$$

from (A-7)-(A-9), the following equations are obtained.

$$\begin{aligned}&J_{11}(s,L)-J_{11}(s,L-1)=-J_{11}(L,L)\{[\ p_{2,2}(L)I_m \quad (1-p_{2,2}(L))I_m\ ][\bar{H}\bar{\Phi}^L J_{11}(s,L-1)\\&+\Phi_V^{L+1}J_{12}(s,L-1)+\Phi_V^L J_{13}(s,L-1)]\\&+[\ p(L)I_m \quad (1-p(L))I_m\ ][\bar{H}\bar{\Phi}^L J_{21}(s,L-1)+\Phi_V^{L+1}J_{22}(s,L-1)+\Phi_V^L J_{23}(s,L-1)]\\&-[\ p_{2,2}(L)I_m \quad (1-p_{2,2}(L))I_m\ ][\bar{H}\bar{\Phi}^L J_{31}(s,L-1)+\Phi_V^{L+1}J_{32}(s,L-1)+\Phi_V^L J_{33}(s,L-1)]\end{aligned}$$

$$\begin{aligned}&J_{12}(s,L)-J_{12}(s,L-1)=-J_{12}(L,L)\{[\ p_{2,2}(L)I_m \quad (1-p_{2,2}(L))I_m\ ][\bar{H}\bar{\Phi}^L J_{11}(s,L-1)\\&+\Phi_V^{L+1}J_{12}(s,L-1)+\Phi_V^L J_{13}(s,L-1)]\\&+[\ p(L)I_m \quad (1-p(L))I_m\ ][\bar{H}\bar{\Phi}^L J_{21}(s,L-1)+\Phi_V^{L+1}J_{22}(s,L-1)+\Phi_V^L J_{23}(s,L-1)]\\&-[\ p_{2,2}(L)I_m \quad (1-p_{2,2}(L))I_m\ ][\bar{H}\bar{\Phi}^L J_{31}(s,L-1)+\Phi_V^{L+1}J_{32}(s,L-1)+\Phi_V^L J_{33}(s,L-1)]\end{aligned}$$

$$\begin{aligned}&J_{13}(s,L)-J_{13}(s,L-1)=-J_{13}(L,L)\{[\ p_{2,2}(L)I_m \quad (1-p_{2,2}(L))I_m\ ][\bar{H}\bar{\Phi}^L J_{11}(s,L-1)\\&+\Phi_V^{L+1}J_{12}(s,L-1)+\Phi_V^L J_{13}(s,L-1)]\\&+[\ p(L)I_m \quad (1-p(L))I_m\ ][\bar{H}\bar{\Phi}^L J_{21}(s,L-1)+\Phi_V^{L+1}J_{22}(s,L-1)+\Phi_V^L J_{23}(s,L-1)]\\&-[\ p_{2,2}(L)I_m \quad (1-p_{2,2}(L))I_m\ ][\bar{H}\bar{\Phi}^L J_{31}(s,L-1)+\Phi_V^{L+1}J_{32}(s,L-1)+\Phi_V^L J_{33}(s,L-1)]\end{aligned}$$

$$\begin{aligned}&J_{21}(s,L)-J_{21}(s,L-1)=-J_{21}(L,L)\{[\ p_{2,2}(L)I_m \quad (1-p_{2,2}(L))I_m\ ][\bar{H}\bar{\Phi}^L J_{11}(s,L-1)\\&+\Phi_V^{L+1}J_{12}(s,L-1)+\Phi_V^L J_{13}(s,L-1)]\\&+[\ p(L)I_m \quad (1-p(L))I_m\ ][\bar{H}\bar{\Phi}^L J_{21}(s,L-1)+\Phi_V^{L+1}J_{22}(s,L-1)+\Phi_V^L J_{23}(s,L-1)]\\&-[\ p_{2,2}(L)I_m \quad (1-p_{2,2}(L))I_m\ ][\bar{H}\bar{\Phi}^L J_{31}(s,L-1)+\Phi_V^{L+1}J_{32}(s,L-1)+\Phi_V^L J_{33}(s,L-1)]\end{aligned}$$

$$\begin{aligned}&J_{22}(s,L)-J_{22}(s,L-1)=-J_{22}(L,L)\{[\ p_{2,2}(L)I_m \quad (1-p_{2,2}(L))I_m\ ][\bar{H}\bar{\Phi}^L J_{11}(s,L-1)\\&+\Phi_V^{L+1}J_{12}(s,L-1)+\Phi_V^L J_{13}(s,L-1)]\\&+[\ p(L)I_m \quad (1-p(L))I_m\ ][\bar{H}\bar{\Phi}^L J_{21}(s,L-1)+\Phi_V^{L+1}J_{22}(s,L-1)+\Phi_V^L J_{23}(s,L-1)]\\&-[\ p_{2,2}(L)I_m \quad (1-p_{2,2}(L))I_m\ ][\bar{H}\bar{\Phi}^L J_{31}(s,L-1)+\Phi_V^{L+1}J_{32}(s,L-1)+\Phi_V^L J_{33}(s,L-1)]\end{aligned}$$

$$\begin{aligned}&J_{23}(s,L)-J_{23}(s,L-1)=-J_{23}(L,L)\{[\ p_{2,2}(L)I_m \quad (1-p_{2,2}(L))I_m\ ][\bar{H}\bar{\Phi}^L J_{11}(s,L-1)\\&+\Phi_V^{L+1}J_{12}(s,L-1)+\Phi_V^L J_{13}(s,L-1)]\\&+[\ p(L)I_m \quad (1-p(L))I_m\ ][\bar{H}\bar{\Phi}^L J_{21}(s,L-1)+\Phi_V^{L+1}J_{22}(s,L-1)+\Phi_V^L J_{23}(s,L-1)]\\&-[\ p_{2,2}(L)I_m \quad (1-p_{2,2}(L))I_m\ ][\bar{H}\bar{\Phi}^L J_{31}(s,L-1)+\Phi_V^{L+1}J_{32}(s,L-1)+\Phi_V^L J_{33}(s,L-1)]\end{aligned}$$

$$\begin{aligned}&J_{31}(s,L)-J_{31}(s,L-1)=-J_{31}(L,L)\{[\ p_{2,2}(L)I_m \quad (1-p_{2,2}(L))I_m\ ][\bar{H}\bar{\Phi}^L J_{11}(s,L-1)\\&+\Phi_V^{L+1}J_{12}(s,L-1)+\Phi_V^L J_{13}(s,L-1)]\\&+[\ p(L)I_m \quad (1-p(L))I_m\ ][\bar{H}\bar{\Phi}^L J_{21}(s,L-1)+\Phi_V^{L+1}J_{22}(s,L-1)+\Phi_V^L J_{23}(s,L-1)]\\&-[\ p_{2,2}(L)I_m \quad (1-p_{2,2}(L))I_m\ ][\bar{H}\bar{\Phi}^L J_{31}(s,L-1)+\Phi_V^{L+1}J_{32}(s,L-1)+\Phi_V^L J_{33}(s,L-1)]\end{aligned}$$

$$\begin{aligned}&J_{32}(s,L)-J_{32}(s,L-1)=-J_{32}(L,L)\{[\ p_{2,2}(L)I_m \quad (1-p_{2,2}(L))I_m\ ][\bar{H}\bar{\Phi}^L J_{11}(s,L-1)\\&+\Phi_V^{L+1}J_{12}(s,L-1)+\Phi_V^L J_{13}(s,L-1)]\\&+[\ p(L)I_m \quad (1-p(L))I_m\ ][\bar{H}\bar{\Phi}^L J_{21}(s,L-1)+\Phi_V^{L+1}J_{22}(s,L-1)+\Phi_V^L J_{23}(s,L-1)]\\&-[\ p_{2,2}(L)I_m \quad (1-p_{2,2}(L))I_m\ ][\bar{H}\bar{\Phi}^L J_{31}(s,L-1)+\Phi_V^{L+1}J_{32}(s,L-1)+\Phi_V^L J_{33}(s,L-1)]\end{aligned}$$

$$J_{33}(s,L)-J_{33}(s,L-1)=-J_{33}(L,L)\{[\ p_{2,2}(L)I_m \quad (1-p_{2,2}(L))I_m\ ][\bar{H}\bar{\Phi}^L J_{11}(s,L-1)$$
$$+\Phi_V^{L+1}J_{12}(s,L-1)+\Phi_V^L J_{13}(s,L-1)]$$
$$+[\ p(L)I_m \quad (1-p(L))I_m\ ][\bar{H}\bar{\Phi}^L J_{21}(s,L-1)+\Phi_V^{L+1}J_{22}(s,L-1)+\Phi_V^L J_{23}(s,L-1)]$$
$$-[\ p_{2,2}(L)I_m \quad (1-p_{2,2}(L))I_m\ ][\bar{H}\bar{\Phi}^L J_{31}(s,L-1)+\Phi_V^{L+1}J_{32}(s,L-1)+\Phi_V^L J_{33}(s,L-1)]$$

Let us express the functions $r_{ij}(L),\quad i=1,..,3,\quad j=1,..,3,$ in the form of their matrix elements as follows:

$$\mathrm{r}_{11}(L)=\begin{bmatrix} r_{1111}(L) & r_{1112}(L) & r_{1113}(L) \\ r_{1121}(L) & r_{1122}(L) & r_{1123}(L) \\ r_{1131}(L) & r_{1132}(L) & r_{1133}(L) \end{bmatrix},\quad r_{12}(L)=\begin{bmatrix} r_{1211}(L) & r_{1212}(L) & r_{1213}(L) \\ r_{1221}(L) & r_{1222}(L) & r_{1223}(L) \\ r_{1231}(L) & r_{1232}(L) & r_{1233}(L) \end{bmatrix},$$

$$r_{13}(L)=\begin{bmatrix} r_{1311}(L) & r_{1312}(L) & r_{1313}(L) \\ r_{1321}(L) & r_{1322}(L) & r_{1323}(L) \\ r_{1331}(L) & r_{1332}(L) & r_{1333}(L) \end{bmatrix},$$

$$r_{21}(L)=\begin{bmatrix} r_{2111}(L) & r_{2112}(L) & r_{2113}(L) \\ r_{2121}(L) & r_{2122}(L) & r_{2123}(L) \\ r_{2131}(L) & r_{2132}(L) & r_{2133}(L) \end{bmatrix},\quad r_{22}(L)=\begin{bmatrix} r_{2211}(L) & r_{2212}(L) & r_{2213}(L) \\ r_{2221}(L) & r_{2222}(L) & r_{2223}(L) \\ r_{2231}(L) & r_{2232}(L) & r_{2233}(L) \end{bmatrix},$$

$$r_{23}(L)=\begin{bmatrix} r_{2311}(L) & r_{2312}(L) & r_{2313}(L) \\ r_{2321}(L) & r_{2322}(L) & r_{2323}(L) \\ r_{2331}(L) & r_{2332}(L) & r_{2333}(L) \end{bmatrix},$$

$$r_{31}(L)=\begin{bmatrix} r_{3111}(L) & r_{3112}(L) & r_{3113}(L) \\ r_{3121}(L) & r_{3122}(L) & r_{3123}(L) \\ r_{3131}(L) & r_{3132}(L) & r_{3133}(L) \end{bmatrix},\quad r_{32}(L)=\begin{bmatrix} r_{3211}(L) & r_{3212}(L) & r_{3213}(L) \\ r_{3221}(L) & r_{3222}(L) & r_{3223}(L) \\ r_{3231}(L) & r_{3232}(L) & r_{3233}(L) \end{bmatrix},$$

$$r_{33}(L)=\begin{bmatrix} r_{3311}(L) & r_{3312}(L) & r_{3313}(L) \\ r_{3321}(L) & r_{3322}(L) & r_{3323}(L) \\ r_{3331}(L) & r_{3332}(L) & r_{3333}(L) \end{bmatrix}.$$

From (A-12)-(A-14), the following equations are obtained.

$$J_{11}(L,L)=\{\bar{\Phi}^{-L}K_{\bar{x}}(L,L)\bar{H}^T\begin{bmatrix} p(L)I_m \\ 0 \end{bmatrix}-[r_{1111}(L-1)(\bar{\Phi}^T)\bar{H}^T+r_{1112}(L-1)(\Phi_V^T)^{L+1}$$
$$+r_{1113}(L-1)(\Phi_V^T)^L]\begin{bmatrix} p_{2,2}(L)I_m \\ (1-p_{2,2}(L))I_m \end{bmatrix}$$
$$-[r_{1211}(L-1)(\bar{\Phi}^T)\bar{H}^T+r_{1212}(L-1)(\Phi_V^T)^{L+1}+r_{1213}(L-1)(\Phi_V^T)^L]\begin{bmatrix} p(L)I_m \\ (1-p(L))I_m \end{bmatrix}$$
$$+[r_{1311}(L-1)(\bar{\Phi}^T)\bar{H}^T+r_{1312}(L-1)(\Phi_V^T)^{L+1}+r_{1313}(L-1)(\Phi_V^T)^L]\begin{bmatrix} p_{2,2}(L)I_m \\ (1-p_{2,2}(L))I_m \end{bmatrix}\}\Gamma^{-1}(L)$$

$$J_{12}(L,L) = \{\Phi_V^{-L} K_V(L,L)\Phi_V^T \begin{bmatrix} p(L)I_m \\ 0 \end{bmatrix} - [r_{1121}(L-1)(\bar{\Phi}^T)\bar{H}^T + r_{1122}(L-1)(\Phi_V^T)^{L+1}$$
$$+ r_{1123}(L-1)(\Phi_V^T)^L] \begin{bmatrix} p_{2,2}(L)I_m \\ (1-p_{2,2}(L))I_m \end{bmatrix}$$
$$-[r_{1221}(L-1)(\bar{\Phi}^T)\bar{H}^T + r_{1222}(L-1)(\Phi_V^T)^{L+1} + r_{1223}(L-1)(\Phi_V^T)^L] \begin{bmatrix} p(L)I_m \\ (1-p(L))I_m \end{bmatrix}$$
$$+[r_{1321}(L-1)(\bar{\Phi}^T)\bar{H}^T + r_{1322}(L-1)(\Phi_V^T)^{L+1} + r_{1323}(L-1)(\Phi_V^T)^L] \begin{bmatrix} p_{2,2}(L)I_m \\ (1-p_{2,2}(L))I_m \end{bmatrix}\}\Gamma^{-1}(L)$$

$$J_{13}(L,L) = \{\Phi_V^{-L} \bar{R}_U(L) \begin{bmatrix} p(L)I_m \\ 0 \end{bmatrix} - [r_{1131}(L-1)(\bar{\Phi}^T)\bar{H}^T + r_{1132}(L-1)(\Phi_V^T)^{L+1}$$
$$+ r_{1133}(L-1)(\Phi_V^T)^L] \begin{bmatrix} p_{2,2}(L)I_m \\ (1-p_{2,2}(L))I_m \end{bmatrix}$$
$$-[r_{1231}(L-1)(\bar{\Phi}^T)\bar{H}^T + r_{1232}(L-1)(\Phi_V^T)^{L+1} + r_{1233}(L-1)(\Phi_V^T)^L] \begin{bmatrix} p(L)I_m \\ (1-p(L))I_m \end{bmatrix}$$
$$+[r_{1331}(L-1)(\bar{\Phi}^T)\bar{H}^T + r_{1332}(L-1)(\Phi_V^T)^{L+1} + r_{1333}(L-1)(\Phi_V^T)^L] \begin{bmatrix} p_{2,2}(L)I_m \\ (1-p_{2,2}(L))I_m \end{bmatrix}\}\Gamma^{-1}(L)$$

$$J_{21}(L,L) = \{\bar{\Phi}^{-L} K_{\bar{x}}(L,L)\bar{H}^T \begin{bmatrix} p(L)I_m \\ 0 \end{bmatrix} - [r_{2111}(L-1)(\bar{\Phi}^T)\bar{H}^T + r_{2112}(L-1)(\Phi_V^T)^{L+1}$$
$$+ r_{2113}(L-1)(\Phi_V^T)^L] \begin{bmatrix} p_{2,2}(L)I_m \\ (1-p_{2,2}(L))I_m \end{bmatrix}$$
$$-[r_{2211}(L-1)(\bar{\Phi}^T)\bar{H}^T + r_{2212}(L-1)(\Phi_V^T)^{L+1} + r_{2213}(L-1)(\Phi_V^T)^L] \begin{bmatrix} p(L)I_m \\ (1-p(L))I_m \end{bmatrix}$$
$$+[r_{2311}(L-1)(\bar{\Phi}^T)\bar{H}^T + r_{2312}(L-1)(\Phi_V^T)^{L+1} + r_{2313}(L-1)(\Phi_V^T)^L] \begin{bmatrix} p_{2,2}(L)I_m \\ (1-p_{2,2}(L))I_m \end{bmatrix}\}\Gamma^{-1}(L)$$

$$J_{22}(L,L) = \{\Phi_V^{-L} K_V(L,L)\Phi_V^T \begin{bmatrix} p(L)I_m \\ 0 \end{bmatrix} - [r_{2121}(L-1)(\bar{\Phi}^T)\bar{H}^T + r_{2122}(L-1)(\Phi_V^T)^{L+1}$$
$$+ r_{2123}(L-1)(\Phi_V^T)^L] \begin{bmatrix} p_{2,2}(L)I_m \\ (1-p_{2,2}(L))I_m \end{bmatrix}$$
$$-[r_{2221}(L-1)(\bar{\Phi}^T)\bar{H}^T + r_{2222}(L-1)(\Phi_V^T)^{L+1} + r_{2223}(L-1)(\Phi_V^T)^L] \begin{bmatrix} p(L)I_m \\ (1-p(L))I_m \end{bmatrix}$$
$$+[r_{2321}(L-1)(\bar{\Phi}^T)\bar{H}^T + r_{2322}(L-1)(\Phi_V^T)^{L+1} + r_{2323}(L-1)(\Phi_V^T)^L] \begin{bmatrix} p_{2,2}(L)I_m \\ (1-p_{2,2}(L))I_m \end{bmatrix}\}\Gamma^{-1}(L)$$

$$J_{23}(L,L) = \{\Phi_V^{-L}\bar{R}_U(L)\begin{bmatrix} p(L)I_m \\ 0 \end{bmatrix} - [r_{2131}(L-1)(\bar{\Phi}^T)\bar{H}^T + r_{2132}(L-1)(\Phi_V^T)^{L+1}$$
$$+ r_{2133}(L-1)(\Phi_V^T)^L]\begin{bmatrix} p_{2,2}(L)I_m \\ (1-p_{2,2}(L))I_m \end{bmatrix}$$
$$-[r_{2231}(L-1)(\bar{\Phi}^T)\bar{H}^T + r_{2232}(L-1)(\Phi_V^T)^{L+1} + r_{2233}(L-1)(\Phi_V^T)^L]\begin{bmatrix} p(L)I_m \\ (1-p(L))I_m \end{bmatrix}$$
$$+[r_{2331}(L-1)(\bar{\Phi}^T)\bar{H}^T + r_{2332}(L-1)(\Phi_V^T)^{L+1} + r_{2333}(L-1)(\Phi_V^T)^L]\begin{bmatrix} p_{2,2}(L)I_m \\ (1-p_{2,2}(L))I_m \end{bmatrix}\}\Gamma^{-1}(L)$$

$$J_{31}(L,L) = \{\bar{\Phi}^{-L}K_{\tilde{x}}(L,L)\bar{H}^T\begin{bmatrix} p(L)I_m \\ 0 \end{bmatrix} - [r_{3111}(L-1)(\bar{\Phi}^T)\bar{H}^T + r_{3112}(L-1)(\Phi_V^T)^{L+1}$$
$$+ r_{3113}(L-1)(\Phi_V^T)^L]\begin{bmatrix} p_{2,2}(L)I_m \\ (1-p_{2,2}(L))I_m \end{bmatrix}$$
$$-[r_{3211}(L-1)(\bar{\Phi}^T)\bar{H}^T + r_{3212}(L-1)(\Phi_V^T)^{L+1} + r_{3213}(L-1)(\Phi_V^T)^L]\begin{bmatrix} p(L)I_m \\ (1-p(L))I_m \end{bmatrix}$$
$$+[r_{3311}(L-1)(\bar{\Phi}^T)\bar{H}^T + r_{3312}(L-1)(\Phi_V^T)^{L+1} + r_{3313}(L-1)(\Phi_V^T)^L]\begin{bmatrix} p_{2,2}(L)I_m \\ (1-p_{2,2}(L))I_m \end{bmatrix}\}\Gamma^{-1}(L)$$

$$J_{32}(L,L) = \{\Phi_V^{-L}K_V(L,L)\Phi_V^T\begin{bmatrix} p(L)I_m \\ 0 \end{bmatrix} - [r_{3121}(L-1)(\bar{\Phi}^T)\bar{H}^T + r_{3122}(L-1)(\Phi_V^T)^{L+1}$$
$$+ r_{3123}(L-1)(\Phi_V^T)^L]\begin{bmatrix} p_{2,2}(L)I_m \\ (1-p_{2,2}(L))I_m \end{bmatrix}$$
$$-[r_{3221}(L-1)(\bar{\Phi}^T)\bar{H}^T + r_{3222}(L-1)(\Phi_V^T)^{L+1} + r_{3223}(L-1)(\Phi_V^T)^L]\begin{bmatrix} p(L)I_m \\ (1-p(L))I_m \end{bmatrix}$$
$$+[r_{3321}(L-1)(\bar{\Phi}^T)\bar{H}^T + r_{3322}(L-1)(\Phi_V^T)^{L+1} + r_{3323}(L-1)(\Phi_V^T)^L]\begin{bmatrix} p_{2,2}(L)I_m \\ (1-p_{2,2}(L))I_m \end{bmatrix}\}\Gamma^{-1}(L)$$

$$J_{33}(L,L) = \{\Phi_V^{-L}\bar{R}_U(L)\begin{bmatrix} p(L)I_m \\ 0 \end{bmatrix} - [r_{3131}(L-1)(\bar{\Phi}^T)\bar{H}^T + r_{3132}(L-1)(\Phi_V^T)^{L+1}$$
$$+ r_{3133}(L-1)(\Phi_V^T)^L]\begin{bmatrix} p_{2,2}(L)I_m \\ (1-p_{2,2}(L))I_m \end{bmatrix}$$
$$-[r_{3231}(L-1)(\bar{\Phi}^T)\bar{H}^T + r_{3232}(L-1)(\Phi_V^T)^{L+1} + r_{3233}(L-1)(\Phi_V^T)^L]\begin{bmatrix} p(L)I_m \\ (1-p(L))I_m \end{bmatrix}$$
$$+[r_{3331}(L-1)(\bar{\Phi}^T)\bar{H}^T + r_{3332}(L-1)(\Phi_V^T)^{L+1} + r_{3333}(L-1)(\Phi_V^T)^L]\begin{bmatrix} p_{2,2}(L)I_m \\ (1-p_{2,2}(L))I_m \end{bmatrix}\}\Gamma^{-1}(L)$$

From (A-15), we have

$$r_{11}(L)=\begin{bmatrix} J_{11}(L,L) \\ J_{12}(L,L) \\ J_{13}(L,L) \end{bmatrix}[\ p(L)I_m \quad 0\ ][\ \bar{H}K_{\bar{x}}(L,L)(\bar{\Phi}^T)^{-L} \quad \Phi_V K_V(L,L)(\Phi_V^T)^{-L} \quad R_U(L)(\Phi_V^T)^{-L}\ ]$$

$$+r_{11}(L-1)-\begin{bmatrix} J_{11}(L,L) \\ J_{12}(L,L) \\ J_{13}(L,L) \end{bmatrix}\{[\ p_{2,2}(L)I_m \quad (1-p_{2,2}(L))I_m\ ][\ \bar{H}\bar{\Phi}^L \quad \Phi_V^{L+1} \quad \Phi_V^L\ ]r_{11}(L-1)$$
$$+[\ p(L)I_m \quad (1-p(L))I_m\ ][\ \bar{H}\bar{\Phi}^L \quad \Phi_V^{L+1} \quad \Phi_V^L\ ]r_{21}(L-1)$$
$$-[\ p_{2,2}(L)I_m \quad (1-p_{2,2}(L))I_m\ ][\ \bar{H}\bar{\Phi}^L \quad \Phi_V^{L+1} \quad \Phi_V^L\ ]r_{31}(L-1)\},\ r_{11}(0)=0.$$

Hence, it is found that $r_{1111}(L)$ is updated by

$$r_{1111}(L)=J_{11}(L,L)[\ p(L)I_m \quad 0\ ]\bar{H}K_{\bar{x}}(L,L)(\bar{\Phi}^T)^{-L}+r_{1111}(L-1)$$
$$-J_{11}(L,L)\{[\ p_{2,2}(L)I_m \quad (1-p_{2,2}(L))I_m\ ][\bar{H}\bar{\Phi}^L r_{1111}(L-1)+\Phi_V^{L+1}r_{1121}(L-1)+\Phi_V^L r_{1131}(L-1)]$$
$$+[\ p(L)I_m \quad (1-p(L))I_m\ ][\bar{H}\bar{\Phi}^L r_{2111}(L-1)+\Phi_V^{L+1}r_{2121}(L-1)+\Phi_V^L r_{2131}(L-1)]$$
$$-[\ p_{2,2}(L)I_m \quad (1-p_{2,2}(L))I_m\ ][\bar{H}\bar{\Phi}^L r_{3111}(L-1)+\Phi_V^{L+1}r_{3121}(L-1)$$
$$+\Phi_V^L r_{3131}(L-1)]\},\quad r_{1111}(0)=0.$$

Similarly, the following equations are obtained.

$$r_{1112}(L)=J_{11}(L,L)[\ p(L)I_m \quad 0\ ]\Phi_V K_V(L,L)(\bar{\Phi}_V^T)^{-L}+r_{1112}(L-1)$$
$$-J_{11}(L,L)\{[\ p_{2,2}(L)I_m \quad (1-p_{2,2}(L))I_m\ ][\bar{H}\bar{\Phi}^L r_{1112}(L-1)+\Phi_V^{L+1}r_{1122}(L-1)+\Phi_V^L r_{1132}(L-1)]$$
$$+[\ p(L)I_m \quad (1-p(L))I_m\ ][\bar{H}\bar{\Phi}^L r_{2112}(L-1)+\Phi_V^{L+1}r_{2122}(L-1)+\Phi_V^L r_{2132}(L-1)]$$
$$-[\ p_{2,2}(L)I_m \quad (1-p_{2,2}(L))I_m\ ][\bar{H}\bar{\Phi}^L r_{3112}(L-1)+\Phi_V^{L+1}r_{3122}(L-1)$$
$$+\Phi_V^L r_{3132}(L-1)]\}\ ,\quad r_{1112}(0)=0$$

$$r_{1113}(L)=J_{11}(L,L)[\ p(L)I_m \quad 0\ ]R_U(L)(\bar{\Phi}_V^T)^{-L}+r_{1113}(L-1)$$
$$-J_{11}(L,L)\{[\ p_{2,2}(L)I_m \quad (1-p_{2,2}(L))I_m\ ][\bar{H}\bar{\Phi}^L r_{1113}(L-1)+\Phi_V^{L+1}r_{1123}(L-1)+\Phi_V^L r_{1133}(L-1)]$$
$$+[\ p(L)I_m \quad (1-p(L))I_m\ ][\bar{H}\bar{\Phi}^L r_{2113}(L-1)+\Phi_V^{L+1}r_{2123}(L-1)+\Phi_V^L r_{2133}(L-1)]$$
$$-[\ p_{2,2}(L)I_m \quad (1-p_{2,2}(L))I_m\ ][\bar{H}\bar{\Phi}^L r_{3113}(L-1)+\Phi_V^{L+1}r_{3123}(L-1)$$
$$+\Phi_V^L r_{3133}(L-1)]\}\ ,\quad r_{1113}(0)=0$$

$$r_{1121}(L)=J_{12}(L,L)[\ p(L)I_m \quad 0\ ]\bar{H}K_{\bar{x}}(L,L)(\bar{\Phi}^T)^{-L}+r_{1121}(L-1)$$
$$-J_{12}(L,L)\{[\ p_{2,2}(L)I_m \quad (1-p_{2,2}(L))I_m\ ][\bar{H}\bar{\Phi}^L r_{1111}(L-1)+\Phi_V^{L+1}r_{1121}(L-1)+\Phi_V^L r_{1131}(L-1)]$$
$$+[\ p(L)I_m \quad (1-p(L))I_m\ ][\bar{H}\bar{\Phi}^L r_{2111}(L-1)+\Phi_V^{L+1}r_{2121}(L-1)+\Phi_V^L r_{2131}(L-1)]$$
$$-[\ p_{2,2}(L)I_m \quad (1-p_{2,2}(L))I_m\ ][\bar{H}\bar{\Phi}^L r_{3111}(L-1)+\Phi_V^{L+1}r_{3121}(L-1)$$
$$+\Phi_V^L r_{3131}(L-1)]\}\ ,\quad r_{1121}(0)=0$$

$$r_{1122}(L) = J_{12}(L,L)[\ p(L)I_m \quad 0\ ]\Phi_V K_V(L,L)(\bar{\Phi}_V^T)^{-L} + r_{1122}(L-1)$$
$$-J_{12}(L,L)\{[\ p_{2,2}(L)I_m \quad (1-p_{2,2}(L))I_m\ ][\bar{H}\bar{\Phi}^L r_{1112}(L-1) + \Phi_V^{L+1} r_{1122}(L-1) + \Phi_V^L r_{1132}(L-1)]$$
$$+[\ p(L)I_m \quad (1-p(L))I_m\ ][\bar{H}\bar{\Phi}^L r_{2112}(L-1) + \Phi_V^{L+1} r_{2122}(L-1) + \Phi_V^L r_{2132}(L-1)]$$
$$-[\ p_{2,2}(L)I_m \quad (1-p_{2,2}(L))I_m\ ][\bar{H}\bar{\Phi}^L r_{3112}(L-1) + \Phi_V^{L+1} r_{3122}(L-1)$$
$$+\Phi_V^L r_{3132}(L-1)]\} \ , \quad r_{1122}(0) = 0$$

$$r_{1123}(L) = J_{12}(L,L)[\ p(L)I_m \quad 0\ ]R_U(L)(\bar{\Phi}_V^T)^{-L} + r_{1123}(L-1)$$
$$-J_{12}(L,L)\{[\ p_{2,2}(L)I_m \quad (1-p_{2,2}(L))I_m\ ][\bar{H}\bar{\Phi}^L r_{1113}(L-1) + \Phi_V^{L+1} r_{1123}(L-1) + \Phi_V^L r_{1133}(L-1)]$$
$$+[\ p(L)I_m \quad (1-p(L))I_m\ ][\bar{H}\bar{\Phi}^L r_{2113}(L-1) + \Phi_V^{L+1} r_{2123}(L-1) + \Phi_V^L r_{2133}(L-1)]$$
$$-[\ p_{2,2}(L)I_m \quad (1-p_{2,2}(L))I_m\ ][\bar{H}\bar{\Phi}^L r_{3113}(L-1) + \Phi_V^{L+1} r_{3123}(L-1)$$
$$+\Phi_V^L r_{3133}(L-1)]\} \ , \quad r_{1123}(0) = 0$$

$$r_{1131}(L) = J_{13}(L,L)[\ p(L)I_m \quad 0\ ]\bar{H}K_{\bar{x}}(L,L)(\bar{\Phi}^T)^{-L} + r_{1131}(L-1)$$
$$-J_{13}(L,L)\{[\ p_{2,2}(L)I_m \quad (1-p_{2,2}(L))I_m\ ][\bar{H}\bar{\Phi}^L r_{1111}(L-1) + \Phi_V^{L+1} r_{1121}(L-1) + \Phi_V^L r_{1131}(L-1)]$$
$$+[\ p(L)I_m \quad (1-p(L))I_m\ ][\bar{H}\bar{\Phi}^L r_{2111}(L-1) + \Phi_V^{L+1} r_{2121}(L-1) + \Phi_V^L r_{2131}(L-1)]$$
$$-[\ p_{2,2}(L)I_m \quad (1-p_{2,2}(L))I_m\ ][\bar{H}\bar{\Phi}^L r_{3111}(L-1) + \Phi_V^{L+1} r_{3121}(L-1)$$
$$+\Phi_V^L r_{3131}(L-1)]\} \ , \quad r_{1131}(0) = 0$$

$$r_{1132}(L) = J_{13}(L,L)[\ p(L)I_m \quad 0\ ]\Phi_V K_V(L,L)(\bar{\Phi}_V^T)^{-L} + r_{1132}(L-1)$$
$$-J_{13}(L,L)\{[\ p_{2,2}(L)I_m \quad (1-p_{2,2}(L))I_m\ ][\bar{H}\bar{\Phi}^L r_{1112}(L-1) + \Phi_V^{L+1} r_{1122}(L-1) + \Phi_V^L r_{1132}(L-1)]$$
$$+[\ p(L)I_m \quad (1-p(L))I_m\ ][\bar{H}\bar{\Phi}^L r_{2112}(L-1) + \Phi_V^{L+1} r_{2122}(L-1) + \Phi_V^L r_{2132}(L-1)]$$
$$-[\ p_{2,2}(L)I_m \quad (1-p_{2,2}(L))I_m\ ][\bar{H}\bar{\Phi}^L r_{3112}(L-1) + \Phi_V^{L+1} r_{3122}(L-1)$$
$$+\Phi_V^L r_{3132}(L-1)]\} \ , \quad r_{1132}(0) = 0$$

$$r_{1133}(L) = J_{13}(L,L)[\ p(L)I_m \quad 0\ ]R_U(L)(\bar{\Phi}_V^T)^{-L} + r_{1133}(L-1)$$
$$-J_{13}(L,L)\{[\ p_{2,2}(L)I_m \quad (1-p_{2,2}(L))I_m\ ][\bar{H}\bar{\Phi}^L r_{1113}(L-1) + \Phi_V^{L+1} r_{1123}(L-1) + \Phi_V^L r_{1133}(L-1)]$$
$$+[\ p(L)I_m \quad (1-p(L))I_m\ ][\bar{H}\bar{\Phi}^L r_{2113}(L-1) + \Phi_V^{L+1} r_{2123}(L-1) + \Phi_V^L r_{2133}(L-1)]$$
$$-[\ p_{2,2}(L)I_m \quad (1-p_{2,2}(L))I_m\ ][\bar{H}\bar{\Phi}^L r_{3113}(L-1) + \Phi_V^{L+1} r_{3123}(L-1)$$
$$+\Phi_V^L r_{3133}(L-1)]\} \ , \quad r_{1133}(0) = 0$$

$$r_{1211}(L) = J_{11}(L,L)[\ 0 \quad I_m\ ]\bar{H}K_{\bar{x}}(L,L)(\bar{\Phi}^T)^{-L} + r_{1211}(L-1)$$
$$-J_{11}(L,L)\{[\ p_{2,2}(L)I_m \quad (1-p_{2,2}(L))I_m\ ][\bar{H}\bar{\Phi}^L r_{1211}(L-1) + \Phi_V^{L+1} r_{1221}(L-1) + \Phi_V^L r_{1231}(L-1)]$$
$$+[\ p(L)I_m \quad (1-p(L))I_m\ ][\bar{H}\bar{\Phi}^L r_{2211}(L-1) + \Phi_V^{L+1} r_{2221}(L-1) + \Phi_V^L r_{2231}(L-1)]$$
$$-[\ p_{2,2}(L)I_m \quad (1-p_{2,2}(L))I_m\ ][\bar{H}\bar{\Phi}^L r_{3211}(L-1) + \Phi_V^{L+1} r_{3221}(L-1)$$
$$+\Phi_V^L r_{3231}(L-1)]\} \ , \quad r_{1211}(0) = 0$$

Similarly, the following equations are obtained.

$$
\begin{aligned}
&r_{1212}(L) = J_{11}(L,L)[\ 0 \quad I_m\ ]\Phi_V K_V(L,L)(\bar{\Phi}_V^T)^{-L} + r_{1212}(L-1)\\
&-J_{11}(L,L)\{[\ p_{2,2}(L)I_m \quad (1-p_{2,2}(L))I_m\ ][\bar{H}\bar{\Phi}^L r_{1212}(L-1) + \Phi_V^{L+1} r_{1222}(L-1) + \Phi_V^L r_{1232}(L-1)]\\
&+[\ p(L)I_m \quad (1-p(L))I_m\ ][\bar{H}\bar{\Phi}^L r_{2212}(L-1) + \Phi_V^{L+1} r_{2222}(L-1) + \Phi_V^L r_{2232}(L-1)]\\
&-[\ p_{2,2}(L)I_m \quad (1-p_{2,2}(L))I_m\ ][\bar{H}\bar{\Phi}^L r_{3212}(L-1) + \Phi_V^{L+1} r_{3222}(L-1)\\
&+\Phi_V^L r_{3232}(L-1)]\}\ , \quad r_{1212}(0) = 0
\end{aligned}
$$

$$
\begin{aligned}
&r_{1213}(L) = J_{11}(L,L)[\ 0 \quad I_m\ ]R_U(L)(\bar{\Phi}_V^T)^{-L} + r_{1213}(L-1)\\
&-J_{11}(L,L)\{[\ p_{2,2}(L)I_m \quad (1-p_{2,2}(L))I_m\ ][\bar{H}\bar{\Phi}^L r_{1213}(L-1) + \Phi_V^{L+1} r_{1223}(L-1) + \Phi_V^L r_{1233}(L-1)]\\
&+[\ p(L)I_m \quad (1-p(L))I_m\ ][\bar{H}\bar{\Phi}^L r_{2213}(L-1) + \Phi_V^{L+1} r_{2223}(L-1) + \Phi_V^L r_{2233}(L-1)]\\
&-[\ p_{2,2}(L)I_m \quad (1-p_{2,2}(L))I_m\ ][\bar{H}\bar{\Phi}^L r_{3213}(L-1) + \Phi_V^{L+1} r_{3223}(L-1)\\
&+\Phi_V^L r_{3233}(L-1)]\}\ , \quad r_{1213}(0) = 0
\end{aligned}
$$

$$
\begin{aligned}
&r_{1221}(L) = J_{12}(L,L)[\ 0 \quad I_m\ ]\bar{H}K_{\bar{x}}(L,L)(\bar{\Phi}^T)^{-L} + r_{1221}(L-1)\\
&-J_{12}(L,L)\{[\ p_{2,2}(L)I_m \quad (1-p_{2,2}(L))I_m\ ][\bar{H}\bar{\Phi}^L r_{1211}(L-1) + \Phi_V^{L+1} r_{1221}(L-1) + \Phi_V^L r_{1231}(L-1)]\\
&+[\ p(L)I_m \quad (1-p(L))I_m\ ][\bar{H}\bar{\Phi}^L r_{2211}(L-1) + \Phi_V^{L+1} r_{2221}(L-1) + \Phi_V^L r_{2231}(L-1)]\\
&-[\ p_{2,2}(L)I_m \quad (1-p_{2,2}(L))I_m\ ][\bar{H}\bar{\Phi}^L r_{3211}(L-1) + \Phi_V^{L+1} r_{3221}(L-1)\\
&+\Phi_V^L r_{3231}(L-1)]\}\ , \quad r_{1221}(0) = 0
\end{aligned}
$$

$$
\begin{aligned}
&r_{1222}(L) = J_{12}(L,L)[\ 0 \quad I_m\ ]\Phi_V K_V(L,L)(\bar{\Phi}_V^T)^{-L} + r_{1222}(L-1)\\
&-J_{12}(L,L)\{[\ p_{2,2}(L)I_m \quad (1-p_{2,2}(L))I_m\ ][\bar{H}\bar{\Phi}^L r_{1212}(L-1) + \Phi_V^{L+1} r_{1222}(L-1) + \Phi_V^L r_{1232}(L-1)]\\
&+[\ p(L)I_m \quad (1-p(L))I_m\ ][\bar{H}\bar{\Phi}^L r_{2212}(L-1) + \Phi_V^{L+1} r_{2222}(L-1) + \Phi_V^L r_{2232}(L-1)]\\
&-[\ p_{2,2}(L)I_m \quad (1-p_{2,2}(L))I_m\ ][\bar{H}\bar{\Phi}^L r_{3212}(L-1) + \Phi_V^{L+1} r_{3222}(L-1)\\
&+\Phi_V^L r_{3232}(L-1)]\}\ , \quad r_{1222}(0) = 0
\end{aligned}
$$

$$
\begin{aligned}
&r_{1223}(L) = J_{12}(L,L)[\ 0 \quad I_m\ ]R_U(L)(\bar{\Phi}_V^T)^{-L} + r_{1223}(L-1)\\
&-J_{12}(L,L)\{[\ p_{2,2}(L)I_m \quad (1-p_{2,2}(L))I_m\ ][\bar{H}\bar{\Phi}^L r_{1213}(L-1) + \Phi_V^{L+1} r_{1223}(L-1) + \Phi_V^L r_{1233}(L-1)]\\
&+[\ p(L)I_m \quad (1-p(L))I_m\ ][\bar{H}\bar{\Phi}^L r_{2213}(L-1) + \Phi_V^{L+1} r_{2223}(L-1) + \Phi_V^L r_{2233}(L-1)]\\
&-[\ p_{2,2}(L)I_m \quad (1-p_{2,2}(L))I_m\ ][\bar{H}\bar{\Phi}^L r_{3213}(L-1) + \Phi_V^{L+1} r_{3223}(L-1)\\
&+\Phi_V^L r_{3233}(L-1)]\}\ , \quad r_{1223}(0) = 0
\end{aligned}
$$

$$
\begin{aligned}
&r_{1231}(L) = J_{13}(L,L)[\ 0 \quad I_m\ ]\bar{H}K_{\bar{x}}(L,L)(\bar{\Phi}^T)^{-L} + r_{1231}(L-1)\\
&-J_{13}(L,L)\{[\ p_{2,2}(L)I_m \quad (1-p_{2,2}(L))I_m\ ][\bar{H}\bar{\Phi}^L r_{1211}(L-1) + \Phi_V^{L+1} r_{1221}(L-1) + \Phi_V^L r_{1231}(L-1)]\\
&+[\ p(L)I_m \quad (1-p(L))I_m\ ][\bar{H}\bar{\Phi}^L r_{2211}(L-1) + \Phi_V^{L+1} r_{2221}(L-1) + \Phi_V^L r_{2231}(L-1)]\\
&-[\ p_{2,2}(L)I_m \quad (1-p_{2,2}(L))I_m\ ][\bar{H}\bar{\Phi}^L r_{3211}(L-1) + \Phi_V^{L+1} r_{3221}(L-1)\\
&+\Phi_V^L r_{3231}(L-1)]\}\ , \quad r_{1231}(0) = 0
\end{aligned}
$$

$$
\begin{aligned}
&r_{1232}(L) = J_{13}(L,L)[\ 0 \quad I_m\ ]\Phi_V K_V(L,L)(\bar{\Phi}_V^T)^{-L} + r_{1232}(L-1)\\
&-J_{13}(L,L)\{[\ p_{2,2}(L)I_m \quad (1-p_{2,2}(L))I_m\ ][\bar{H}\bar{\Phi}^L r_{1212}(L-1) + \Phi_V^{L+1} r_{1222}(L-1) + \Phi_V^L r_{1232}(L-1)]\\
&+[\ p(L)I_m \quad (1-p(L))I_m\ ][\bar{H}\bar{\Phi}^L r_{2212}(L-1) + \Phi_V^{L+1} r_{2222}(L-1) + \Phi_V^L r_{2232}(L-1)]\\
&-[\ p_{2,2}(L)I_m \quad (1-p_{2,2}(L))I_m\ ][\bar{H}\bar{\Phi}^L r_{3212}(L-1) + \Phi_V^{L+1} r_{3222}(L-1)\\
&+\Phi_V^L r_{3232}(L-1)]\}\ , \quad r_{1232}(0) = 0
\end{aligned}
$$

$$r_{1233}(L)=J_{13}(L,L)[\ 0 \quad I_m\ ]R_U(L)(\bar{\Phi}_V^T)^{-L}+r_{1233}(L-1)$$
$$-J_{13}(L,L)\{[\ p_{2,2}(L)I_m \quad (1-p_{2,2}(L))I_m\ ][\bar{H}\bar{\Phi}^L r_{1213}(L-1)+\Phi_V^{L+1}r_{1223}(L-1)+\Phi_V^L r_{1233}(L-1)]$$
$$+[\ p(L)I_m \quad (1-p(L))I_m\ ][\bar{H}\bar{\Phi}^L r_{2213}(L-1)+\Phi_V^{L+1}r_{2223}(L-1)+\Phi_V^L r_{2233}(L-1)]$$
$$-[\ p_{2,2}(L)I_m \quad (1-p_{2,2}(L))I_m\ ][\bar{H}\bar{\Phi}^L r_{3213}(L-1)+\Phi_V^{L+1}r_{3223}(L-1)$$
$$+\Phi_V^L r_{3233}(L-1)]\} \ , \quad r_{1233}(0)=0$$

$$r_{1311}(L)=J_{11}(L,L)[\ 0 \quad p(L)I_m\ ]\bar{H}K_{\tilde{x}}(L,L)(\bar{\Phi}^T)^{-L}+r_{1311}(L-1)$$
$$-J_{11}(L,L)\{[\ p_{2,2}(L)I_m \quad (1-p_{2,2}(L))I_m\ ][\bar{H}\bar{\Phi}^L r_{1311}(L-1)+\Phi_V^{L+1}r_{1321}(L-1)+\Phi_V^L r_{1331}(L-1)]$$
$$+[\ p(L)I_m \quad (1-p(L))I_m\ ][\bar{H}\bar{\Phi}^L r_{2311}(L-1)+\Phi_V^{L+1}r_{2321}(L-1)+\Phi_V^L r_{2331}(L-1)]$$
$$-[\ p_{2,2}(L)I_m \quad (1-p_{2,2}(L))I_m\ ][\bar{H}\bar{\Phi}^L r_{3311}(L-1)+\Phi_V^{L+1}r_{3321}(L-1)$$
$$+\Phi_V^L r_{3331}(L-1)]\} \ , \quad r_{1311}(0)=0$$

$$r_{1312}(L)=J_{11}(L,L)[\ 0 \quad p(L)I_m\ ]\Phi_V K_V(L,L)(\bar{\Phi}_V^T)^{-L}+r_{1312}(L-1)$$
$$-J_{11}(L,L)\{[\ p_{2,2}(L)I_m \quad (1-p_{2,2}(L))I_m\ ][\bar{H}\bar{\Phi}^L r_{1312}(L-1)+\Phi_V^{L+1}r_{1322}(L-1)+\Phi_V^L r_{1332}(L-1)]$$
$$+[\ p(L)I_m \quad (1-p(L))I_m\ ][\bar{H}\bar{\Phi}^L r_{2312}(L-1)+\Phi_V^{L+1}r_{2322}(L-1)+\Phi_V^L r_{2332}(L-1)]$$
$$-[\ p_{2,2}(L)I_m \quad (1-p_{2,2}(L))I_m\ ][\bar{H}\bar{\Phi}^L r_{3312}(L-1)+\Phi_V^{L+1}r_{3322}(L-1)$$
$$+\Phi_V^L r_{3332}(L-1)]\} \ , \quad r_{1312}(0)=0$$

$$r_{1313}(L)=J_{11}(L,L)[\ 0 \quad p(L)I_m\ ]R_U(L)(\bar{\Phi}_V^T)^{-L}+r_{1313}(L-1)$$
$$-J_{11}(L,L)\{[\ p_{2,2}(L)I_m \quad (1-p_{2,2}(L))I_m\ ][\bar{H}\bar{\Phi}^L r_{1313}(L-1)+\Phi_V^{L+1}r_{1323}(L-1)+\Phi_V^L r_{1333}(L-1)]$$
$$+[\ p(L)I_m \quad (1-p(L))I_m\ ][\bar{H}\bar{\Phi}^L r_{2313}(L-1)+\Phi_V^{L+1}r_{2323}(L-1)+\Phi_V^L r_{2333}(L-1)]$$
$$-[\ p_{2,2}(L)I_m \quad (1-p_{2,2}(L))I_m\ ][\bar{H}\bar{\Phi}^L r_{3313}(L-1)+\Phi_V^{L+1}r_{3323}(L-1)$$
$$+\Phi_V^L r_{3333}(L-1)]\} \ , \quad r_{1313}(0)=0$$

$$r_{1321}(L)=J_{12}(L,L)[\ 0 \quad p(L)I_m\ ]\bar{H}K_{\tilde{x}}(L,L)(\bar{\Phi}^T)^{-L}+r_{1321}(L-1)$$
$$-J_{12}(L,L)\{[\ p_{2,2}(L)I_m \quad (1-p_{2,2}(L))I_m\ ][\bar{H}\bar{\Phi}^L r_{1311}(L-1)+\Phi_V^{L+1}r_{1321}(L-1)+\Phi_V^L r_{1331}(L-1)]$$
$$+[\ p(L)I_m \quad (1-p(L))I_m\ ][\bar{H}\bar{\Phi}^L r_{2311}(L-1)+\Phi_V^{L+1}r_{2321}(L-1)+\Phi_V^L r_{2331}(L-1)]$$
$$-[\ p_{2,2}(L)I_m \quad (1-p_{2,2}(L))I_m\ ][\bar{H}\bar{\Phi}^L r_{3311}(L-1)+\Phi_V^{L+1}r_{3321}(L-1)$$
$$+\Phi_V^L r_{3331}(L-1)]\} \ , \quad r_{1321}(0)=0$$

$$r_{1322}(L)=J_{12}(L,L)[\ 0 \quad p(L)I_m\ ]\Phi_V K_V(L,L)(\bar{\Phi}_V^T)^{-L}+r_{1322}(L-1)$$
$$-J_{12}(L,L)\{[\ p_{2,2}(L)I_m \quad (1-p_{2,2}(L))I_m\ ][\bar{H}\bar{\Phi}^L r_{1312}(L-1)+\Phi_V^{L+1}r_{1322}(L-1)+\Phi_V^L r_{1332}(L-1)]$$
$$+[\ p(L)I_m \quad (1-p(L))I_m\ ][\bar{H}\bar{\Phi}^L r_{2312}(L-1)+\Phi_V^{L+1}r_{2322}(L-1)+\Phi_V^L r_{2332}(L-1)]$$
$$-[\ p_{2,2}(L)I_m \quad (1-p_{2,2}(L))I_m\ ][\bar{H}\bar{\Phi}^L r_{3312}(L-1)+\Phi_V^{L+1}r_{3322}(L-1)$$
$$+\Phi_V^L r_{3332}(L-1)]\} \ , \quad r_{1322}(0)=0$$

$$r_{1323}(L)=J_{12}(L,L)[\ 0 \quad p(L)I_m\ ]R_U(L)(\bar{\Phi}_V^T)^{-L}+r_{1323}(L-1)$$
$$-J_{12}(L,L)\{[\ p_{2,2}(L)I_m \quad (1-p_{2,2}(L))I_m\ ][\bar{H}\bar{\Phi}^L r_{1313}(L-1)+\Phi_V^{L+1}r_{1323}(L-1)+\Phi_V^L r_{1333}(L-1)]$$
$$+[\ p(L)I_m \quad (1-p(L))I_m\ ][\bar{H}\bar{\Phi}^L r_{2313}(L-1)+\Phi_V^{L+1}r_{2323}(L-1)+\Phi_V^L r_{2333}(L-1)]$$
$$-[\ p_{2,2}(L)I_m \quad (1-p_{2,2}(L))I_m\ ][\bar{H}\bar{\Phi}^L r_{3313}(L-1)+\Phi_V^{L+1}r_{3323}(L-1)$$
$$+\Phi_V^L r_{3333}(L-1)]\} \ , \quad r_{1323}(0)=0$$

$$r_{1331}(L) = J_{13}(L,L)[\ 0 \quad p(L)I_m\ ]\bar{H}K_{\bar{x}}(L,L)(\bar{\Phi}^T)^{-L} + r_{1331}(L-1)$$
$$-J_{13}(L,L)\{[\ p_{2,2}(L)I_m \quad (1-p_{2,2}(L))I_m\ ][\bar{H}\bar{\Phi}^L r_{1311}(L-1) + \Phi_V^{L+1} r_{1321}(L-1) + \Phi_V^L r_{1331}(L-1)]$$
$$+[\ p(L)I_m \quad (1-p(L))I_m\ ][\bar{H}\bar{\Phi}^L r_{2311}(L-1) + \Phi_V^{L+1} r_{2321}(L-1) + \Phi_V^L r_{2331}(L-1)]$$
$$-[\ p_{2,2}(L)I_m \quad (1-p_{2,2}(L))I_m\ ][\bar{H}\bar{\Phi}^L r_{3311}(L-1) + \Phi_V^{L+1} r_{3321}(L-1)$$
$$+\Phi_V^L r_{3331}(L-1)]\}\ , \quad r_{1331}(0) = 0$$

$$r_{1332}(L) = J_{13}(L,L)[\ 0 \quad p(L)I_m\ ]\Phi_V K_V(L,L)(\bar{\Phi}_V^T)^{-L} + r_{1332}(L-1)$$
$$-J_{13}(L,L)\{[\ p_{2,2}(L)I_m \quad (1-p_{2,2}(L))I_m\ ][\bar{H}\bar{\Phi}^L r_{1312}(L-1) + \Phi_V^{L+1} r_{1322}(L-1) + \Phi_V^L r_{1332}(L-1)]$$
$$+[\ p(L)I_m \quad (1-p(L))I_m\ ][\bar{H}\bar{\Phi}^L r_{2312}(L-1) + \Phi_V^{L+1} r_{2322}(L-1) + \Phi_V^L r_{2332}(L-1)]$$
$$-[\ p_{2,2}(L)I_m \quad (1-p_{2,2}(L))I_m\ ][\bar{H}\bar{\Phi}^L r_{3312}(L-1) + \Phi_V^{L+1} r_{3322}(L-1)$$
$$+\Phi_V^L r_{3332}(L-1)]\}\ , \quad r_{1332}(0) = 0$$

$$r_{1333}(L) = J_{13}(L,L)[\ 0 \quad p(L)I_m\ ]R_U(L)(\bar{\Phi}_V^T)^{-L} + r_{1333}(L-1)$$
$$-J_{13}(L,L)\{[\ p_{2,2}(L)I_m \quad (1-p_{2,2}(L))I_m\ ][\bar{H}\bar{\Phi}^L r_{1313}(L-1) + \Phi_V^{L+1} r_{1323}(L-1) + \Phi_V^L r_{1333}(L-1)]$$
$$+[\ p(L)I_m \quad (1-p(L))I_m\ ][\bar{H}\bar{\Phi}^L r_{2313}(L-1) + \Phi_V^{L+1} r_{2323}(L-1) + \Phi_V^L r_{2333}(L-1)]$$
$$-[\ p_{2,2}(L)I_m \quad (1-p_{2,2}(L))I_m\ ][\bar{H}\bar{\Phi}^L r_{3313}(L-1) + \Phi_V^{L+1} r_{3323}(L-1)$$
$$+\Phi_V^L r_{3333}(L-1)]\}\ , \quad r_{1333}(0) = 0$$

$$r_{2111}(L) = J_{21}(L,L)[\ p(L)I_m \quad 0\ ]\bar{H}K_{\bar{x}}(L,L)(\bar{\Phi}^T)^{-L} + r_{2111}(L-1)$$
$$-J_{21}(L,L)\{[\ p_{2,2}(L)I_m \quad (1-p_{2,2}(L))I_m\ ][\bar{H}\bar{\Phi}^L r_{1111}(L-1) + \Phi_V^{L+1} r_{1121}(L-1) + \Phi_V^L r_{1131}(L-1)]$$
$$+[\ p(L)I_m \quad (1-p(L))I_m\ ][\bar{H}\bar{\Phi}^L r_{2111}(L-1) + \Phi_V^{L+1} r_{2121}(L-1) + \Phi_V^L r_{2131}(L-1)]$$
$$-[\ p_{2,2}(L)I_m \quad (1-p_{2,2}(L))I_m\ ][\bar{H}\bar{\Phi}^L r_{3111}(L-1) + \Phi_V^{L+1} r_{3121}(L-1)$$
$$+\Phi_V^L r_{3131}(L-1)]\}\ , \quad r_{2111}(0) = 0$$

$$r_{2112}(L) = J_{21}(L,L)[\ p(L)I_m \quad 0\ ]\Phi_V K_V(L,L)(\bar{\Phi}_V^T)^{-L} + r_{2112}(L-1)$$
$$-J_{21}(L,L)\{[\ p_{2,2}(L)I_m \quad (1-p_{2,2}(L))I_m\ ][\bar{H}\bar{\Phi}^L r_{1112}(L-1) + \Phi_V^{L+1} r_{1122}(L-1) + \Phi_V^L r_{1132}(L-1)]$$
$$+[\ p(L)I_m \quad (1-p(L))I_m\ ][\bar{H}\bar{\Phi}^L r_{2112}(L-1) + \Phi_V^{L+1} r_{2122}(L-1) + \Phi_V^L r_{2132}(L-1)]$$
$$-[\ p_{2,2}(L)I_m \quad (1-p_{2,2}(L))I_m\ ][\bar{H}\bar{\Phi}^L r_{3112}(L-1) + \Phi_V^{L+1} r_{3122}(L-1)$$
$$+\Phi_V^L r_{3132}(L-1)]\}\ , \quad r_{2112}(0) = 0$$

$$r_{2113}(L) = J_{21}(L,L)[\ p(L)I_m \quad 0\ ]R_U(L)(\bar{\Phi}_V^T)^{-L} + r_{2113}(L-1)$$
$$-J_{21}(L,L)\{[\ p_{2,2}(L)I_m \quad (1-p_{2,2}(L))I_m\ ][\bar{H}\bar{\Phi}^L r_{1113}(L-1) + \Phi_V^{L+1} r_{1123}(L-1) + \Phi_V^L r_{1133}(L-1)]$$
$$+[\ p(L)I_m \quad (1-p(L))I_m\ ][\bar{H}\bar{\Phi}^L r_{2113}(L-1) + \Phi_V^{L+1} r_{2123}(L-1) + \Phi_V^L r_{2133}(L-1)]$$
$$-[\ p_{2,2}(L)I_m \quad (1-p_{2,2}(L))I_m\ ][\bar{H}\bar{\Phi}^L r_{3113}(L-1) + \Phi_V^{L+1} r_{3123}(L-1)$$
$$+\Phi_V^L r_{3133}(L-1)]\}\ , \quad r_{2113}(0) = 0$$

$$r_{2121}(L) = J_{22}(L,L)[\ p(L)I_m \quad 0\ ]\bar{H}K_{\bar{x}}(L,L)(\bar{\Phi}^T)^{-L} + r_{2121}(L-1)$$
$$-J_{22}(L,L)\{[\ p_{2,2}(L)I_m \quad (1-p_{2,2}(L))I_m\ ][\bar{H}\bar{\Phi}^L r_{1111}(L-1) + \Phi_V^{L+1} r_{1121}(L-1) + \Phi_V^L r_{1131}(L-1)]$$
$$+[\ p(L)I_m \quad (1-p(L))I_m\ ][\bar{H}\bar{\Phi}^L r_{2111}(L-1) + \Phi_V^{L+1} r_{2121}(L-1) + \Phi_V^L r_{2131}(L-1)]$$
$$-[\ p_{2,2}(L)I_m \quad (1-p_{2,2}(L))I_m\ ][\bar{H}\bar{\Phi}^L r_{3111}(L-1) + \Phi_V^{L+1} r_{3121}(L-1)$$
$$+\Phi_V^L r_{3131}(L-1)]\}\ , \quad r_{2121}(0) = 0$$

$$
\begin{aligned}
&r_{2122}(L) = J_{22}(L,L)[\ p(L)I_m \quad 0\ ]\Phi_V K_V(L,L)(\bar{\Phi}_V^T)^{-L} + r_{2122}(L-1)\\
&-J_{22}(L,L)\{[\ p_{2,2}(L)I_m \quad (1-p_{2,2}(L))I_m\ ][\bar{H}\bar{\Phi}^L r_{1112}(L-1) + \Phi_V^{L+1} r_{1122}(L-1) + \Phi_V^L r_{1132}(L-1)]\\
&+[\ p(L)I_m \quad (1-p(L))I_m\ ][\bar{H}\bar{\Phi}^L r_{2112}(L-1) + \Phi_V^{L+1} r_{2122}(L-1) + \Phi_V^L r_{2132}(L-1)]\\
&-[\ p_{2,2}(L)I_m \quad (1-p_{2,2}(L))I_m\ ][\bar{H}\bar{\Phi}^L r_{3112}(L-1) + \Phi_V^{L+1} r_{3122}(L-1)\\
&+\Phi_V^L r_{3132}(L-1)]\} \ , \quad r_{2122}(0) = 0
\end{aligned}
$$

$$
\begin{aligned}
&r_{2123}(L) = J_{22}(L,L)[\ p(L)I_m \quad 0\ ]R_U(L)(\bar{\Phi}_V^T)^{-L} + r_{2123}(L-1)\\
&-J_{22}(L,L)\{[\ p_{2,2}(L)I_m \quad (1-p_{2,2}(L))I_m\ ][\bar{H}\bar{\Phi}^L r_{1113}(L-1) + \Phi_V^{L+1} r_{1123}(L-1) + \Phi_V^L r_{1133}(L-1)]\\
&+[\ p(L)I_m \quad (1-p(L))I_m\ ][\bar{H}\bar{\Phi}^L r_{2113}(L-1) + \Phi_V^{L+1} r_{2123}(L-1) + \Phi_V^L r_{2133}(L-1)]\\
&-[\ p_{2,2}(L)I_m \quad (1-p_{2,2}(L))I_m\ ][\bar{H}\bar{\Phi}^L r_{3113}(L-1) + \Phi_V^{L+1} r_{3123}(L-1)\\
&+\Phi_V^L r_{3133}(L-1)]\} \ , \quad r_{2123}(0) = 0
\end{aligned}
$$

$$
\begin{aligned}
&r_{2131}(L) = J_{23}(L,L)[\ p(L)I_m \quad 0\ ]\bar{H}K_{\bar{x}}(L,L)(\bar{\Phi}^T)^{-L} + r_{2131}(L-1)\\
&-J_{23}(L,L)\{[\ p_{2,2}(L)I_m \quad (1-p_{2,2}(L))I_m\ ][\bar{H}\bar{\Phi}^L r_{1111}(L-1) + \Phi_V^{L+1} r_{1121}(L-1) + \Phi_V^L r_{1131}(L-1)]\\
&+[\ p(L)I_m \quad (1-p(L))I_m\ ][\bar{H}\bar{\Phi}^L r_{2111}(L-1) + \Phi_V^{L+1} r_{2121}(L-1) + \Phi_V^L r_{2131}(L-1)]\\
&-[\ p_{2,2}(L)I_m \quad (1-p_{2,2}(L))I_m\ ][\bar{H}\bar{\Phi}^L r_{3111}(L-1) + \Phi_V^{L+1} r_{3121}(L-1)\\
&+\Phi_V^L r_{3131}(L-1)]\} \ , \quad r_{2131}(0) = 0
\end{aligned}
$$

$$
\begin{aligned}
&r_{2132}(L) = J_{23}(L,L)[\ p(L)I_m \quad 0\ ]\Phi_V K_V(L,L)(\bar{\Phi}_V^T)^{-L} + r_{2132}(L-1)\\
&-J_{23}(L,L)\{[\ p_{2,2}(L)I_m \quad (1-p_{2,2}(L))I_m\ ][\bar{H}\bar{\Phi}^L r_{1112}(L-1) + \Phi_V^{L+1} r_{1122}(L-1) + \Phi_V^L r_{1132}(L-1)]\\
&+[\ p(L)I_m \quad (1-p(L))I_m\ ][\bar{H}\bar{\Phi}^L r_{2112}(L-1) + \Phi_V^{L+1} r_{2122}(L-1) + \Phi_V^L r_{2132}(L-1)]\\
&-[\ p_{2,2}(L)I_m \quad (1-p_{2,2}(L))I_m\ ][\bar{H}\bar{\Phi}^L r_{3112}(L-1) + \Phi_V^{L+1} r_{3122}(L-1)\\
&+\Phi_V^L r_{3132}(L-1)]\} \ , \quad r_{2132}(0) = 0
\end{aligned}
$$

$$
\begin{aligned}
&r_{2133}(L) = J_{23}(L,L)[\ p(L)I_m \quad 0\ ]R_U(L)(\bar{\Phi}_V^T)^{-L} + r_{2133}(L-1)\\
&-J_{23}(L,L)\{[\ p_{2,2}(L)I_m \quad (1-p_{2,2}(L))I_m\ ][\bar{H}\bar{\Phi}^L r_{1113}(L-1) + \Phi_V^{L+1} r_{1123}(L-1) + \Phi_V^L r_{1133}(L-1)]\\
&+[\ p(L)I_m \quad (1-p(L))I_m\ ][\bar{H}\bar{\Phi}^L r_{2113}(L-1) + \Phi_V^{L+1} r_{2123}(L-1) + \Phi_V^L r_{2133}(L-1)]\\
&-[\ p_{2,2}(L)I_m \quad (1-p_{2,2}(L))I_m\ ][\bar{H}\bar{\Phi}^L r_{3113}(L-1) + \Phi_V^{L+1} r_{3123}(L-1)\\
&+\Phi_V^L r_{3133}(L-1)]\} \ , \quad r_{2133}(0) = 0
\end{aligned}
$$

$$
\begin{aligned}
&r_{2211}(L) = J_{11}(L,L)[\ 0 \quad I_m\ ]\bar{H}K_{\bar{x}}(L,L)(\bar{\Phi}^T)^{-L} + r_{2211}(L-1)\\
&-J_{21}(L,L)\{[\ p_{2,2}(L)I_m \quad (1-p_{2,2}(L))I_m\ ][\bar{H}\bar{\Phi}^L r_{1211}(L-1) + \Phi_V^{L+1} r_{1221}(L-1) + \Phi_V^L r_{1231}(L-1)]\\
&+[\ p(L)I_m \quad (1-p(L))I_m\ ][\bar{H}\bar{\Phi}^L r_{2211}(L-1) + \Phi_V^{L+1} r_{2221}(L-1) + \Phi_V^L r_{2231}(L-1)]\\
&-[\ p_{2,2}(L)I_m \quad (1-p_{2,2}(L))I_m\ ][\bar{H}\bar{\Phi}^L r_{3211}(L-1) + \Phi_V^{L+1} r_{3221}(L-1)\\
&+\Phi_V^L r_{3231}(L-1)]\} \ , \quad r_{2211}(0) = 0
\end{aligned}
$$

$$
\begin{aligned}
&r_{2212}(L) = J_{21}(L,L)[\ 0 \quad I_m\ ]\Phi_V K_V(L,L)(\bar{\Phi}_V^T)^{-L} + r_{2212}(L-1)\\
&-J_{21}(L,L)\{[\ p_{2,2}(L)I_m \quad (1-p_{2,2}(L))I_m\ ][\bar{H}\bar{\Phi}^L r_{1212}(L-1) + \Phi_V^{L+1} r_{1222}(L-1) + \Phi_V^L r_{1232}(L-1)]\\
&+[\ p(L)I_m \quad (1-p(L))I_m\ ][\bar{H}\bar{\Phi}^L r_{2212}(L-1) + \Phi_V^{L+1} r_{2222}(L-1) + \Phi_V^L r_{2232}(L-1)]\\
&-[\ p_{2,2}(L)I_m \quad (1-p_{2,2}(L))I_m\ ][\bar{H}\bar{\Phi}^L r_{3212}(L-1) + \Phi_V^{L+1} r_{3222}(L-1)\\
&+\Phi_V^L r_{3232}(L-1)]\} \ , \quad r_{2212}(0) = 0
\end{aligned}
$$

$$r_{2213}(L)=J_{21}(L,L)[\ 0\quad I_m\ ]R_U(L)(\bar{\Phi}_V^T)^{-L}+r_{2213}(L-1)$$
$$-J_{21}(L,L)\{[\ p_{2,2}(L)I_m\quad (1-p_{2,2}(L))I_m\ ][\bar{H}\bar{\Phi}^L r_{1213}(L-1)+\Phi_V^{L+1}r_{1223}(L-1)+\Phi_V^L r_{1233}(L-1)]$$
$$+[\ p(L)I_m\quad (1-p(L))I_m\ ][\bar{H}\bar{\Phi}^L r_{2213}(L-1)+\Phi_V^{L+1}r_{2223}(L-1)+\Phi_V^L r_{2233}(L-1)]$$
$$-[\ p_{2,2}(L)I_m\quad (1-p_{2,2}(L))I_m\ ][\bar{H}\bar{\Phi}^L r_{3213}(L-1)+\Phi_V^{L+1}r_{3223}(L-1)$$
$$+\Phi_V^L r_{3233}(L-1)]\},\qquad r_{2213}(0)=0$$

$$r_{2221}(L)=J_{22}(L,L)[\ 0\quad I_m\ ]\bar{H}K_{\bar{x}}(L,L)(\bar{\Phi}^T)^{-L}+r_{2221}(L-1)$$
$$-J_{22}(L,L)\{[\ p_{2,2}(L)I_m\quad (1-p_{2,2}(L))I_m\ ][\bar{H}\bar{\Phi}^L r_{1211}(L-1)+\Phi_V^{L+1}r_{1221}(L-1)+\Phi_V^L r_{1231}(L-1)]$$
$$+[\ p(L)I_m\quad (1-p(L))I_m\ ][\bar{H}\bar{\Phi}^L r_{2211}(L-1)+\Phi_V^{L+1}r_{2221}(L-1)+\Phi_V^L r_{2231}(L-1)]$$
$$-[\ p_{2,2}(L)I_m\quad (1-p_{2,2}(L))I_m\ ][\bar{H}\bar{\Phi}^L r_{3211}(L-1)+\Phi_V^{L+1}r_{3221}(L-1)$$
$$+\Phi_V^L r_{3231}(L-1)]\},\qquad r_{2221}(0)=0$$

$$r_{2222}(L)=J_{22}(L,L)[\ 0\quad I_m\ ]\Phi_V K_V(L,L)(\bar{\Phi}_V^T)^{-L}+r_{2222}(L-1)$$
$$-J_{22}(L,L)\{[\ p_{2,2}(L)I_m\quad (1-p_{2,2}(L))I_m\ ][\bar{H}\bar{\Phi}^L r_{1212}(L-1)+\Phi_V^{L+1}r_{1222}(L-1)+\Phi_V^L r_{1232}(L-1)]$$
$$+[\ p(L)I_m\quad (1-p(L))I_m\ ][\bar{H}\bar{\Phi}^L r_{2212}(L-1)+\Phi_V^{L+1}r_{2222}(L-1)+\Phi_V^L r_{2232}(L-1)]$$
$$-[\ p_{2,2}(L)I_m\quad (1-p_{2,2}(L))I_m\ ][\bar{H}\bar{\Phi}^L r_{3212}(L-1)+\Phi_V^{L+1}r_{3222}(L-1)$$
$$+\Phi_V^L r_{3232}(L-1)]\},\qquad r_{2222}(0)=0$$

$$r_{2223}(L)=J_{22}(L,L)[\ 0\quad I_m\ ]R_U(L)(\bar{\Phi}_V^T)^{-L}+r_{2223}(L-1)$$
$$-J_{22}(L,L)\{[\ p_{2,2}(L)I_m\quad (1-p_{2,2}(L))I_m\ ][\bar{H}\bar{\Phi}^L r_{1213}(L-1)+\Phi_V^{L+1}r_{1223}(L-1)+\Phi_V^L r_{1233}(L-1)]$$
$$+[\ p(L)I_m\quad (1-p(L))I_m\ ][\bar{H}\bar{\Phi}^L r_{2213}(L-1)+\Phi_V^{L+1}r_{2223}(L-1)+\Phi_V^L r_{2233}(L-1)]$$
$$-[\ p_{2,2}(L)I_m\quad (1-p_{2,2}(L))I_m\ ][\bar{H}\bar{\Phi}^L r_{3213}(L-1)+\Phi_V^{L+1}r_{3223}(L-1)$$
$$+\Phi_V^L r_{3233}(L-1)]\},\qquad r_{2223}(0)=0$$

$$r_{2231}(L)=J_{23}(L,L)[\ 0\quad I_m\ ]\bar{H}K_{\bar{x}}(L,L)(\bar{\Phi}^T)^{-L}+r_{2231}(L-1)$$
$$-J_{23}(L,L)\{[\ p_{2,2}(L)I_m\quad (1-p_{2,2}(L))I_m\ ][\bar{H}\bar{\Phi}^L r_{1211}(L-1)+\Phi_V^{L+1}r_{1221}(L-1)+\Phi_V^L r_{1231}(L-1)]$$
$$+[\ p(L)I_m\quad (1-p(L))I_m\ ][\bar{H}\bar{\Phi}^L r_{2211}(L-1)+\Phi_V^{L+1}r_{2221}(L-1)+\Phi_V^L r_{2231}(L-1)]$$
$$-[\ p_{2,2}(L)I_m\quad (1-p_{2,2}(L))I_m\ ][\bar{H}\bar{\Phi}^L r_{3211}(L-1)+\Phi_V^{L+1}r_{3221}(L-1)$$
$$+\Phi_V^L r_{3231}(L-1)]\},\qquad r_{2231}(0)=0$$

$$r_{2232}(L)=J_{23}(L,L)[\ 0\quad I_m\ ]\Phi_V K_V(L,L)(\bar{\Phi}_V^T)^{-L}+r_{2232}(L-1)$$
$$-J_{23}(L,L)\{[\ p_{2,2}(L)I_m\quad (1-p_{2,2}(L))I_m\ ][\bar{H}\bar{\Phi}^L r_{1212}(L-1)+\Phi_V^{L+1}r_{1222}(L-1)+\Phi_V^L r_{1232}(L-1)]$$
$$+[\ p(L)I_m\quad (1-p(L))I_m\ ][\bar{H}\bar{\Phi}^L r_{2212}(L-1)+\Phi_V^{L+1}r_{2222}(L-1)+\Phi_V^L r_{2232}(L-1)]$$
$$-[\ p_{2,2}(L)I_m\quad (1-p_{2,2}(L))I_m\ ][\bar{H}\bar{\Phi}^L r_{3212}(L-1)+\Phi_V^{L+1}r_{3222}(L-1)$$
$$+\Phi_V^L r_{3232}(L-1)]\},\qquad r_{2232}(0)=0$$

$$r_{2233}(L)=J_{23}(L,L)[\ 0\quad I_m\ ]R_U(L)(\bar{\Phi}_V^T)^{-L}+r_{2233}(L-1)$$
$$-J_{23}(L,L)\{[\ p_{2,2}(L)I_m\quad (1-p_{2,2}(L))I_m\ ][\bar{H}\bar{\Phi}^L r_{1213}(L-1)+\Phi_V^{L+1}r_{1223}(L-1)+\Phi_V^L r_{1233}(L-1)]$$
$$+[\ p(L)I_m\quad (1-p(L))I_m\ ][\bar{H}\bar{\Phi}^L r_{2213}(L-1)+\Phi_V^{L+1}r_{2223}(L-1)+\Phi_V^L r_{2233}(L-1)]$$
$$-[\ p_{2,2}(L)I_m\quad (1-p_{2,2}(L))I_m\ ][\bar{H}\bar{\Phi}^L r_{3213}(L-1)+\Phi_V^{L+1}r_{3223}(L-1)$$
$$+\Phi_V^L r_{3233}(L-1)]\},\qquad r_{2233}(0)=0$$

$$r_{2311}(L) = J_{21}(L,L)[\ 0 \quad p(L)I_m\ ]\bar{H}K_{\bar{x}}(L,L)(\bar{\Phi}^T)^{-L} + r_{2311}(L-1)$$
$$-J_{21}(L,L)\{[\ p_{2,2}(L)I_m \quad (1-p_{2,2}(L))I_m\ ][\bar{H}\bar{\Phi}^L r_{1311}(L-1) + \Phi_V^{L+1} r_{1321}(L-1) + \Phi_V^L r_{1331}(L-1)]$$
$$+[\ p(L)I_m \quad (1-p(L))I_m\ ][\bar{H}\bar{\Phi}^L r_{2311}(L-1) + \Phi_V^{L+1} r_{2321}(L-1) + \Phi_V^L r_{2331}(L-1)]$$
$$-[\ p_{2,2}(L)I_m \quad (1-p_{2,2}(L))I_m\ ][\bar{H}\bar{\Phi}^L r_{3311}(L-1) + \Phi_V^{L+1} r_{3321}(L-1)$$
$$+\Phi_V^L r_{3331}(L-1)]\}, \quad r_{2311}(0) = 0$$

$$r_{2312}(L) = J_{21}(L,L)[\ 0 \quad p(L)I_m\ ]\Phi_V K_V(L,L)(\bar{\Phi}_V^T)^{-L} + r_{2312}(L-1)$$
$$-J_{21}(L,L)\{[\ p_{2,2}(L)I_m \quad (1-p_{2,2}(L))I_m\ ][\bar{H}\bar{\Phi}^L r_{1312}(L-1) + \Phi_V^{L+1} r_{1322}(L-1) + \Phi_V^L r_{1332}(L-1)]$$
$$+[\ p(L)I_m \quad (1-p(L))I_m\ ][\bar{H}\bar{\Phi}^L r_{2312}(L-1) + \Phi_V^{L+1} r_{2322}(L-1) + \Phi_V^L r_{2332}(L-1)]$$
$$-[\ p_{2,2}(L)I_m \quad (1-p_{2,2}(L))I_m\ ][\bar{H}\bar{\Phi}^L r_{3312}(L-1) + \Phi_V^{L+1} r_{3322}(L-1)$$
$$+\Phi_V^L r_{3332}(L-1)]\}, \quad r_{2312}(0) = 0$$

$$r_{2313}(L) = J_{21}(L,L)[\ 0 \quad p(L)I_m\ ]R_U(L)(\bar{\Phi}_V^T)^{-L} + r_{2313}(L-1)$$
$$-J_{21}(L,L)\{[\ p_{2,2}(L)I_m \quad (1-p_{2,2}(L))I_m\ ][\bar{H}\bar{\Phi}^L r_{1313}(L-1) + \Phi_V^{L+1} r_{1323}(L-1) + \Phi_V^L r_{1333}(L-1)]$$
$$+[\ p(L)I_m \quad (1-p(L))I_m\ ][\bar{H}\bar{\Phi}^L r_{2313}(L-1) + \Phi_V^{L+1} r_{2323}(L-1) + \Phi_V^L r_{2333}(L-1)]$$
$$-[\ p_{2,2}(L)I_m \quad (1-p_{2,2}(L))I_m\ ][\bar{H}\bar{\Phi}^L r_{3313}(L-1) + \Phi_V^{L+1} r_{3323}(L-1)$$
$$+\Phi_V^L r_{3333}(L-1)]\}, \quad r_{2313}(0) = 0$$

$$r_{2321}(L) = J_{22}(L,L)[\ 0 \quad p(L)I_m\ ]\bar{H}K_{\bar{x}}(L,L)(\bar{\Phi}^T)^{-L} + r_{2321}(L-1)$$
$$-J_{22}(L,L)\{[\ p_{2,2}(L)I_m \quad (1-p_{2,2}(L))I_m\ ][\bar{H}\bar{\Phi}^L r_{1311}(L-1) + \Phi_V^{L+1} r_{1321}(L-1) + \Phi_V^L r_{1331}(L-1)]$$
$$+[\ p(L)I_m \quad (1-p(L))I_m\ ][\bar{H}\bar{\Phi}^L r_{2311}(L-1) + \Phi_V^{L+1} r_{2321}(L-1) + \Phi_V^L r_{2331}(L-1)]$$
$$-[\ p_{2,2}(L)I_m \quad (1-p_{2,2}(L))I_m\ ][\bar{H}\bar{\Phi}^L r_{3311}(L-1) + \Phi_V^{L+1} r_{3321}(L-1)$$
$$+\Phi_V^L r_{3331}(L-1)]\}, \quad r_{2321}(0) = 0$$

$$r_{2322}(L) = J_{22}(L,L)[\ 0 \quad p(L)I_m\ ]\Phi_V K_V(L,L)(\bar{\Phi}_V^T)^{-L} + r_{2322}(L-1)$$
$$-J_{22}(L,L)\{[\ p_{2,2}(L)I_m \quad (1-p_{2,2}(L))I_m\ ][\bar{H}\bar{\Phi}^L r_{1312}(L-1) + \Phi_V^{L+1} r_{1322}(L-1) + \Phi_V^L r_{1332}(L-1)]$$
$$+[\ p(L)I_m \quad (1-p(L))I_m\ ][\bar{H}\bar{\Phi}^L r_{2312}(L-1) + \Phi_V^{L+1} r_{2322}(L-1) + \Phi_V^L r_{2332}(L-1)]$$
$$-[\ p_{2,2}(L)I_m \quad (1-p_{2,2}(L))I_m\ ][\bar{H}\bar{\Phi}^L r_{3312}(L-1) + \Phi_V^{L+1} r_{3322}(L-1)$$
$$+\Phi_V^L r_{3332}(L-1)]\}, \quad r_{1322}(0) = 0$$

$$r_{2323}(L) = J_{22}(L,L)[\ 0 \quad p(L)I_m\ ]R_U(L)(\bar{\Phi}_V^T)^{-L} + r_{2323}(L-1)$$
$$-J_{22}(L,L)\{[\ p_{2,2}(L)I_m \quad (1-p_{2,2}(L))I_m\ ][\bar{H}\bar{\Phi}^L r_{1313}(L-1) + \Phi_V^{L+1} r_{1323}(L-1) + \Phi_V^L r_{1333}(L-1)]$$
$$+[\ p(L)I_m \quad (1-p(L))I_m\ ][\bar{H}\bar{\Phi}^L r_{2313}(L-1) + \Phi_V^{L+1} r_{2323}(L-1) + \Phi_V^L r_{2333}(L-1)]$$
$$-[\ p_{2,2}(L)I_m \quad (1-p_{2,2}(L))I_m\ ][\bar{H}\bar{\Phi}^L r_{3313}(L-1) + \Phi_V^{L+1} r_{3323}(L-1)$$
$$+\Phi_V^L r_{3333}(L-1)]\}, \quad r_{2323}(0) = 0$$

$$r_{2331}(L) = J_{23}(L,L)[\ 0 \quad p(L)I_m\ ]\bar{H}K_{\bar{x}}(L,L)(\bar{\Phi}^T)^{-L} + r_{2331}(L-1)$$
$$-J_{23}(L,L)\{[\ p_{2,2}(L)I_m \quad (1-p_{2,2}(L))I_m\ ][\bar{H}\bar{\Phi}^L r_{1311}(L-1) + \Phi_V^{L+1} r_{1321}(L-1) + \Phi_V^L r_{1331}(L-1)]$$
$$+[\ p(L)I_m \quad (1-p(L))I_m\ ][\bar{H}\bar{\Phi}^L r_{2311}(L-1) + \Phi_V^{L+1} r_{2321}(L-1) + \Phi_V^L r_{2331}(L-1)]$$
$$-[\ p_{2,2}(L)I_m \quad (1-p_{2,2}(L))I_m\ ][\bar{H}\bar{\Phi}^L r_{3311}(L-1) + \Phi_V^{L+1} r_{3321}(L-1)$$
$$+\Phi_V^L r_{3331}(L-1)]\}, \quad r_{2331}(0) = 0$$

$$r_{2332}(L) = J_{23}(L,L)[\ 0 \quad p(L)I_m\ ]\Phi_V K_V(L,L)(\bar{\Phi}_V^T)^{-L} + r_{2332}(L-1)$$
$$-J_{23}(L,L)\{[\ p_{2,2}(L)I_m \quad (1-p_{2,2}(L))I_m\ ][\bar{H}\bar{\Phi}^L r_{1312}(L-1) + \Phi_V^{L+1} r_{1322}(L-1) + \Phi_V^L r_{1332}(L-1)]$$
$$+[\ p(L)I_m \quad (1-p(L))I_m\ ][\bar{H}\bar{\Phi}^L r_{2312}(L-1) + \Phi_V^{L+1} r_{2322}(L-1) + \Phi_V^L r_{2332}(L-1)]$$
$$-[\ p_{2,2}(L)I_m \quad (1-p_{2,2}(L))I_m\ ][\bar{H}\bar{\Phi}^L r_{3312}(L-1) + \Phi_V^{L+1} r_{3322}(L-1)$$
$$+\Phi_V^L r_{3332}(L-1)]\}\ , \quad r_{2332}(0) = 0$$

$$r_{2333}(L) = J_{23}(L,L)[\ 0 \quad p(L)I_m\ ]R_U(L)(\bar{\Phi}_V^T)^{-L} + r_{2333}(L-1)$$
$$-J_{23}(L,L)\{[\ p_{2,2}(L)I_m \quad (1-p_{2,2}(L))I_m\ ][\bar{H}\bar{\Phi}^L r_{1313}(L-1) + \Phi_V^{L+1} r_{1323}(L-1) + \Phi_V^L r_{1333}(L-1)]$$
$$+[\ p(L)I_m \quad (1-p(L))I_m\ ][\bar{H}\bar{\Phi}^L r_{2313}(L-1) + \Phi_V^{L+1} r_{2323}(L-1) + \Phi_V^L r_{2333}(L-1)]$$
$$-[\ p_{2,2}(L)I_m \quad (1-p_{2,2}(L))I_m\ ][\bar{H}\bar{\Phi}^L r_{3313}(L-1) + \Phi_V^{L+1} r_{3323}(L-1)$$
$$+\Phi_V^L r_{3333}(L-1)]\}, \quad r_{2333}(0) = 0$$

$$r_{3111}(L) = J_{31}(L,L)[\ p(L)I_m \quad 0\ ]\bar{H}K_{\bar{x}}(L,L)(\bar{\Phi}^T)^{-L} + r_{3111}(L-1)$$
$$-J_{31}(L,L)\{[\ p_{2,2}(L)I_m \quad (1-p_{2,2}(L))I_m\ ][\bar{H}\bar{\Phi}^L r_{1111}(L-1) + \Phi_V^{L+1} r_{1121}(L-1) + \Phi_V^L r_{1131}(L-1)]$$
$$+[\ p(L)I_m \quad (1-p(L))I_m\ ][\bar{H}\bar{\Phi}^L r_{2111}(L-1) + \Phi_V^{L+1} r_{2121}(L-1) + \Phi_V^L r_{2131}(L-1)]$$
$$-[\ p_{2,2}(L)I_m \quad (1-p_{2,2}(L))I_m\ ][\bar{H}\bar{\Phi}^L r_{3111}(L-1) + \Phi_V^{L+1} r_{3121}(L-1)$$
$$+\Phi_V^L r_{3131}(L-1)]\}, \quad r_{3111}(0) = 0$$

$$r_{3112}(L) = J_{31}(L,L)[\ p(L)I_m \quad 0\ ]\Phi_V K_V(L,L)(\bar{\Phi}_V^T)^{-L} + r_{3112}(L-1)$$
$$-J_{31}(L,L)\{[\ p_{2,2}(L)I_m \quad (1-p_{2,2}(L))I_m\ ][\bar{H}\bar{\Phi}^L r_{1112}(L-1) + \Phi_V^{L+1} r_{1122}(L-1) + \Phi_V^L r_{1132}(L-1)]$$
$$+[\ p(L)I_m \quad (1-p(L))I_m\ ][\bar{H}\bar{\Phi}^L r_{2112}(L-1) + \Phi_V^{L+1} r_{2122}(L-1) + \Phi_V^L r_{2132}(L-1)]$$
$$-[\ p_{2,2}(L)I_m \quad (1-p_{2,2}(L))I_m\ ][\bar{H}\bar{\Phi}^L r_{3112}(L-1) + \Phi_V^{L+1} r_{3122}(L-1)$$
$$+\Phi_V^L r_{3132}(L-1)]\}, \quad r_{3112}(0) = 0$$

$$r_{3113}(L) = J_{31}(L,L)[\ p(L)I_m \quad 0\ ]R_U(L)(\bar{\Phi}_V^T)^{-L} + r_{3113}(L-1)$$
$$-J_{31}(L,L)\{[\ p_{2,2}(L)I_m \quad (1-p_{2,2}(L))I_m\ ][\bar{H}\bar{\Phi}^L r_{1113}(L-1) + \Phi_V^{L+1} r_{1123}(L-1) + \Phi_V^L r_{1133}(L-1)]$$
$$+[\ p(L)I_m \quad (1-p(L))I_m\ ][\bar{H}\bar{\Phi}^L r_{2113}(L-1) + \Phi_V^{L+1} r_{2123}(L-1) + \Phi_V^L r_{2133}(L-1)]$$
$$-[\ p_{2,2}(L)I_m \quad (1-p_{2,2}(L))I_m\ ][\bar{H}\bar{\Phi}^L r_{3113}(L-1) + \Phi_V^{L+1} r_{3123}(L-1)$$
$$+\Phi_V^L r_{3133}(L-1)]\}\ , \quad r_{3113}(0) = 0$$

$$r_{3121}(L) = J_{32}(L,L)[\ p(L)I_m \quad 0\ ]\bar{H}K_{\bar{x}}(L,L)(\bar{\Phi}^T)^{-L} + r_{3121}(L-1)$$
$$-J_{32}(L,L)\{[\ p_{2,2}(L)I_m \quad (1-p_{2,2}(L))I_m\ ][\bar{H}\bar{\Phi}^L r_{1111}(L-1) + \Phi_V^{L+1} r_{1121}(L-1) + \Phi_V^L r_{1131}(L-1)]$$
$$+[\ p(L)I_m \quad (1-p(L))I_m\ ][\bar{H}\bar{\Phi}^L r_{2111}(L-1) + \Phi_V^{L+1} r_{2121}(L-1) + \Phi_V^L r_{2131}(L-1)]$$
$$-[\ p_{2,2}(L)I_m \quad (1-p_{2,2}(L))I_m\ ][\bar{H}\bar{\Phi}^L r_{3111}(L-1) + \Phi_V^{L+1} r_{3121}(L-1)$$
$$+\Phi_V^L r_{3131}(L-1)]\}\ , \quad r_{3121}(0) = 0$$

$$r_{3122}(L) = J_{32}(L,L)[\ p(L)I_m \quad 0\ ]\Phi_V K_V(L,L)(\bar{\Phi}_V^T)^{-L} + r_{3122}(L-1)$$
$$-J_{32}(L,L)\{[\ p_{2,2}(L)I_m \quad (1-p_{2,2}(L))I_m\ ][\bar{H}\bar{\Phi}^L r_{1112}(L-1) + \Phi_V^{L+1} r_{1122}(L-1) + \Phi_V^L r_{1132}(L-1)]$$
$$+[\ p(L)I_m \quad (1-p(L))I_m\ ][\bar{H}\bar{\Phi}^L r_{2112}(L-1) + \Phi_V^{L+1} r_{2122}(L-1) + \Phi_V^L r_{2132}(L-1)]$$
$$-[\ p_{2,2}(L)I_m \quad (1-p_{2,2}(L))I_m\ ][\bar{H}\bar{\Phi}^L r_{3112}(L-1) + \Phi_V^{L+1} r_{3122}(L-1)$$
$$+\Phi_V^L r_{3132}(L-1)]\}\ , \quad r_{3122}(0) = 0$$

$$r_{3123}(L) = J_{32}(L,L)[\ p(L)I_m \quad 0\ ]R_U(L)(\bar{\Phi}_V^T)^{-L} + r_{3123}(L-1)$$
$$-J_{32}(L,L)\{[\ p_{2,2}(L)I_m \quad (1-p_{2,2}(L))I_m\ ][\bar{H}\bar{\Phi}^L r_{1113}(L-1) + \Phi_V^{L+1} r_{1123}(L-1) + \Phi_V^L r_{1133}(L-1)]$$
$$+[\ p(L)I_m \quad (1-p(L))I_m\ ][\bar{H}\bar{\Phi}^L r_{2113}(L-1) + \Phi_V^{L+1} r_{2123}(L-1) + \Phi_V^L r_{2133}(L-1)]$$
$$-[\ p_{2,2}(L)I_m \quad (1-p_{2,2}(L))I_m\ ][\bar{H}\bar{\Phi}^L r_{3113}(L-1) + \Phi_V^{L+1} r_{3123}(L-1)$$
$$+\Phi_V^L r_{3133}(L-1)]\} \ , \quad r_{3123}(0) = 0$$

$$r_{3131}(L) = J_{33}(L,L)[\ p(L)I_m \quad 0\ ]\bar{H}K_{\bar{x}}(L,L)(\bar{\Phi}^T)^{-L} + r_{3131}(L-1)$$
$$-J_{33}(L,L)\{[\ p_{2,2}(L)I_m \quad (1-p_{2,2}(L))I_m\ ][\bar{H}\bar{\Phi}^L r_{1111}(L-1) + \Phi_V^{L+1} r_{1121}(L-1) + \Phi_V^L r_{1131}(L-1)]$$
$$+[\ p(L)I_m \quad (1-p(L))I_m\ ][\bar{H}\bar{\Phi}^L r_{2111}(L-1) + \Phi_V^{L+1} r_{2121}(L-1) + \Phi_V^L r_{2131}(L-1)]$$
$$-[\ p_{2,2}(L)I_m \quad (1-p_{2,2}(L))I_m\ ][\bar{H}\bar{\Phi}^L r_{3111}(L-1) + \Phi_V^{L+1} r_{3121}(L-1)$$
$$+\Phi_V^L r_{3131}(L-1)]\} \ , \quad r_{3131}(0) = 0$$

$$r_{3132}(L) = J_{33}(L,L)[\ p(L)I_m \quad 0\ ]\Phi_V K_V(L,L)(\bar{\Phi}_V^T)^{-L} + r_{3132}(L-1)$$
$$-J_{33}(L,L)\{[\ p_{2,2}(L)I_m \quad (1-p_{2,2}(L))I_m\ ][\bar{H}\bar{\Phi}^L r_{1112}(L-1) + \Phi_V^{L+1} r_{1122}(L-1) + \Phi_V^L r_{1132}(L-1)]$$
$$+[\ p(L)I_m \quad (1-p(L))I_m\ ][\bar{H}\bar{\Phi}^L r_{2112}(L-1) + \Phi_V^{L+1} r_{2122}(L-1) + \Phi_V^L r_{2132}(L-1)]$$
$$-[\ p_{2,2}(L)I_m \quad (1-p_{2,2}(L))I_m\ ][\bar{H}\bar{\Phi}^L r_{3112}(L-1) + \Phi_V^{L+1} r_{3122}(L-1)$$
$$+\Phi_V^L r_{3132}(L-1)]\} \ , \quad r_{3132}(0) = 0$$

$$r_{3133}(L) = J_{33}(L,L)[\ p(L)I_m \quad 0\ ]R_U(L)(\bar{\Phi}_V^T)^{-L} + r_{3133}(L-1)$$
$$-J_{23}(L,L)\{[\ p_{2,2}(L)I_m \quad (1-p_{2,2}(L))I_m\ ][\bar{H}\bar{\Phi}^L r_{1113}(L-1) + \Phi_V^{L+1} r_{1123}(L-1) + \Phi_V^L r_{1133}(L-1)]$$
$$+[\ p(L)I_m \quad (1-p(L))I_m\ ][\bar{H}\bar{\Phi}^L r_{2113}(L-1) + \Phi_V^{L+1} r_{2123}(L-1) + \Phi_V^L r_{2133}(L-1)]$$
$$-[\ p_{2,2}(L)I_m \quad (1-p_{2,2}(L))I_m\ ][\bar{H}\bar{\Phi}^L r_{3113}(L-1) + \Phi_V^{L+1} r_{3123}(L-1)$$
$$+\Phi_V^L r_{3133}(L-1)]\} \ , \quad r_{3133}(0) = 0$$

$$r_{3211}(L) = J_{31}(L,L)[\ 0 \quad I_m\ ]\bar{H}K_{\bar{x}}(L,L)(\bar{\Phi}^T)^{-L} + r_{3211}(L-1)$$
$$-J_{31}(L,L)\{[\ p_{2,2}(L)I_m \quad (1-p_{2,2}(L))I_m\ ][\bar{H}\bar{\Phi}^L r_{1211}(L-1) + \Phi_V^{L+1} r_{1221}(L-1) + \Phi_V^L r_{1231}(L-1)]$$
$$+[\ p(L)I_m \quad (1-p(L))I_m\ ][\bar{H}\bar{\Phi}^L r_{2211}(L-1) + \Phi_V^{L+1} r_{2221}(L-1) + \Phi_V^L r_{2231}(L-1)]$$
$$-[\ p_{2,2}(L)I_m \quad (1-p_{2,2}(L))I_m\ ][\bar{H}\bar{\Phi}^L r_{3211}(L-1) + \Phi_V^{L+1} r_{3221}(L-1)$$
$$+\Phi_V^L r_{3231}(L-1)]\} \ , \quad r_{3211}(0) = 0$$

$$r_{3212}(L) = J_{31}(L,L)[\ 0 \quad I_m\ ]\Phi_V K_V(L,L)(\bar{\Phi}_V^T)^{-L} + r_{3212}(L-1)$$
$$-J_{31}(L,L)\{[\ p_{2,2}(L)I_m \quad (1-p_{2,2}(L))I_m\ ][\bar{H}\bar{\Phi}^L r_{1212}(L-1) + \Phi_V^{L+1} r_{1222}(L-1) + \Phi_V^L r_{1232}(L-1)]$$
$$+[\ p(L)I_m \quad (1-p(L))I_m\ ][\bar{H}\bar{\Phi}^L r_{2212}(L-1) + \Phi_V^{L+1} r_{2222}(L-1) + \Phi_V^L r_{2232}(L-1)]$$
$$-[\ p_{2,2}(L)I_m \quad (1-p_{2,2}(L))I_m\ ][\bar{H}\bar{\Phi}^L r_{3212}(L-1) + \Phi_V^{L+1} r_{3222}(L-1)$$
$$+\Phi_V^L r_{3232}(L-1)]\} \ , \quad r_{3212}(0) = 0$$

$$r_{3213}(L) = J_{31}(L,L)[\ 0 \quad I_m\ ]R_U(L)(\bar{\Phi}_V^T)^{-L} + r_{3213}(L-1)$$
$$-J_{31}(L,L)\{[\ p_{2,2}(L)I_m \quad (1-p_{2,2}(L))I_m\ ][\bar{H}\bar{\Phi}^L r_{1213}(L-1) + \Phi_V^{L+1} r_{1223}(L-1) + \Phi_V^L r_{1233}(L-1)]$$
$$+[\ p(L)I_m \quad (1-p(L))I_m\ ][\bar{H}\bar{\Phi}^L r_{2213}(L-1) + \Phi_V^{L+1} r_{2223}(L-1) + \Phi_V^L r_{2233}(L-1)]$$
$$-[\ p_{2,2}(L)I_m \quad (1-p_{2,2}(L))I_m\ ][\bar{H}\bar{\Phi}^L r_{3213}(L-1) + \Phi_V^{L+1} r_{3223}(L-1)$$
$$+\Phi_V^L r_{3233}(L-1)]\} \ , \quad r_{3213}(0) = 0$$

$$r_{3221}(L) = J_{32}(L,L)[\ 0 \quad I_m\ ]\bar{H}K_{\bar{x}}(L,L)(\bar{\Phi}^T)^{-L} + r_{3221}(L-1)$$
$$-J_{32}(L,L)\{[\ p_{2,2}(L)I_m \quad (1-p_{2,2}(L))I_m\ ][\bar{H}\bar{\Phi}^L r_{1211}(L-1) + \Phi_V^{L+1} r_{1221}(L-1) + \Phi_V^L r_{1231}(L-1)]$$
$$+[\ p(L)I_m \quad (1-p(L))I_m\ ][\bar{H}\bar{\Phi}^L r_{2211}(L-1) + \Phi_V^{L+1} r_{2221}(L-1) + \Phi_V^L r_{2231}(L-1)]$$
$$-[\ p_{2,2}(L)I_m \quad (1-p_{2,2}(L))I_m\ ][\bar{H}\bar{\Phi}^L r_{3211}(L-1) + \Phi_V^{L+1} r_{3221}(L-1)$$
$$+\Phi_V^L r_{3231}(L-1)]\}\ , \quad r_{3221}(0) = 0$$

$$r_{3222}(L) = J_{32}(L,L)[\ 0 \quad I_m\ ]\Phi_V K_V(L,L)(\bar{\Phi}_V^T)^{-L} + r_{3222}(L-1)$$
$$-J_{32}(L,L)\{[\ p_{2,2}(L)I_m \quad (1-p_{2,2}(L))I_m\ ][\bar{H}\bar{\Phi}^L r_{1212}(L-1) + \Phi_V^{L+1} r_{1222}(L-1) + \Phi_V^L r_{1232}(L-1)]$$
$$+[\ p(L)I_m \quad (1-p(L))I_m\ ][\bar{H}\bar{\Phi}^L r_{2212}(L-1) + \Phi_V^{L+1} r_{2222}(L-1) + \Phi_V^L r_{2232}(L-1)]$$
$$-[\ p_{2,2}(L)I_m \quad (1-p_{2,2}(L))I_m\ ][\bar{H}\bar{\Phi}^L r_{3212}(L-1) + \Phi_V^{L+1} r_{3222}(L-1)$$
$$+\Phi_V^L r_{3232}(L-1)]\}\ , \quad r_{3222}(0) = 0$$

$$r_{3223}(L) = J_{32}(L,L)[\ 0 \quad I_m\ ]R_U(L)(\bar{\Phi}_V^T)^{-L} + r_{3223}(L-1)$$
$$-J_{32}(L,L)\{[\ p_{2,2}(L)I_m \quad (1-p_{2,2}(L))I_m\ ][\bar{H}\bar{\Phi}^L r_{1213}(L-1) + \Phi_V^{L+1} r_{1223}(L-1) + \Phi_V^L r_{1233}(L-1)]$$
$$+[\ p(L)I_m \quad (1-p(L))I_m\ ][\bar{H}\bar{\Phi}^L r_{2213}(L-1) + \Phi_V^{L+1} r_{2223}(L-1) + \Phi_V^L r_{2233}(L-1)]$$
$$-[\ p_{2,2}(L)I_m \quad (1-p_{2,2}(L))I_m\ ][\bar{H}\bar{\Phi}^L r_{3213}(L-1) + \Phi_V^{L+1} r_{3223}(L-1)$$
$$+\Phi_V^L r_{3233}(L-1)]\}\ , \quad r_{3223}(0) = 0$$

$$r_{3231}(L) = J_{33}(L,L)[\ 0 \quad I_m\ ]\bar{H}K_{\bar{x}}(L,L)(\bar{\Phi}^T)^{-L} + r_{3231}(L-1)$$
$$-J_{33}(L,L)\{[\ p_{2,2}(L)I_m \quad (1-p_{2,2}(L))I_m\ ][\bar{H}\bar{\Phi}^L r_{1211}(L-1) + \Phi_V^{L+1} r_{1221}(L-1) + \Phi_V^L r_{1231}(L-1)]$$
$$+[\ p(L)I_m \quad (1-p(L))I_m\ ][\bar{H}\bar{\Phi}^L r_{2211}(L-1) + \Phi_V^{L+1} r_{2221}(L-1) + \Phi_V^L r_{2231}(L-1)]$$
$$-[\ p_{2,2}(L)I_m \quad (1-p_{2,2}(L))I_m\ ][\bar{H}\bar{\Phi}^L r_{3211}(L-1) + \Phi_V^{L+1} r_{3221}(L-1)$$
$$+\Phi_V^L r_{3231}(L-1)]\}\ , \quad r_{3231}(0) = 0$$

$$r_{3232}(L) = J_{33}(L,L)[\ 0 \quad I_m\ ]\Phi_V K_V(L,L)(\bar{\Phi}_V^T)^{-L} + r_{3232}(L-1)$$
$$-J_{33}(L,L)\{[\ p_{2,2}(L)I_m \quad (1-p_{2,2}(L))I_m\ ][\bar{H}\bar{\Phi}^L r_{1212}(L-1) + \Phi_V^{L+1} r_{1222}(L-1) + \Phi_V^L r_{1232}(L-1)]$$
$$+[\ p(L)I_m \quad (1-p(L))I_m\ ][\bar{H}\bar{\Phi}^L r_{2212}(L-1) + \Phi_V^{L+1} r_{2222}(L-1) + \Phi_V^L r_{2232}(L-1)]$$
$$-[\ p_{2,2}(L)I_m \quad (1-p_{2,2}(L))I_m\ ][\bar{H}\bar{\Phi}^L r_{3212}(L-1) + \Phi_V^{L+1} r_{3222}(L-1)$$
$$+\Phi_V^L r_{3232}(L-1)]\}\ , \quad r_{3232}(0) = 0$$

$$r_{3233}(L) = J_{33}(L,L)[\ 0 \quad I_m\ ]R_U(L)(\bar{\Phi}_V^T)^{-L} + r_{3233}(L-1)$$
$$-J_{33}(L,L)\{[\ p_{2,2}(L)I_m \quad (1-p_{2,2}(L))I_m\ ][\bar{H}\bar{\Phi}^L r_{1213}(L-1) + \Phi_V^{L+1} r_{1223}(L-1) + \Phi_V^L r_{1233}(L-1)]$$
$$+[\ p(L)I_m \quad (1-p(L))I_m\ ][\bar{H}\bar{\Phi}^L r_{2213}(L-1) + \Phi_V^{L+1} r_{2223}(L-1) + \Phi_V^L r_{2233}(L-1)]$$
$$-[\ p_{2,2}(L)I_m \quad (1-p_{2,2}(L))I_m\ ][\bar{H}\bar{\Phi}^L r_{3213}(L-1) + \Phi_V^{L+1} r_{3223}(L-1)$$
$$+\Phi_V^L r_{3233}(L-1)]\}\ , \quad r_{3233}(0) = 0$$

$$r_{3311}(L) = J_{31}(L,L)[\ 0 \quad p(L)I_m\ ]\bar{H}K_{\bar{x}}(L,L)(\bar{\Phi}^T)^{-L} + r_{3311}(L-1)$$
$$-J_{31}(L,L)\{[\ p_{2,2}(L)I_m \quad (1-p_{2,2}(L))I_m\ ][\bar{H}\bar{\Phi}^L r_{1311}(L-1) + \Phi_V^{L+1} r_{1321}(L-1) + \Phi_V^L r_{1331}(L-1)]$$
$$+[\ p(L)I_m \quad (1-p(L))I_m\ ][\bar{H}\bar{\Phi}^L r_{2311}(L-1) + \Phi_V^{L+1} r_{2321}(L-1) + \Phi_V^L r_{2331}(L-1)]$$
$$-[\ p_{2,2}(L)I_m \quad (1-p_{2,2}(L))I_m\ ][\bar{H}\bar{\Phi}^L r_{3311}(L-1) + \Phi_V^{L+1} r_{3321}(L-1)$$
$$+\Phi_V^L r_{3331}(L-1)]\}\ , \quad r_{3311}(0) = 0$$

$$r_{3312}(L) = J_{31}(L,L)[\ 0 \quad p(L)I_m\ ]\Phi_V K_V(L,L)(\bar{\Phi}_V^T)^{-L} + r_{3312}(L-1)$$
$$-J_{31}(L,L)\{[\ p_{2,2}(L)I_m \quad (1-p_{2,2}(L))I_m\ ][\bar{H}\bar{\Phi}^L r_{1312}(L-1) + \Phi_V^{L+1} r_{1322}(L-1) + \Phi_V^L r_{1332}(L-1)]$$
$$+[\ p(L)I_m \quad (1-p(L))I_m\ ][\bar{H}\bar{\Phi}^L r_{2312}(L-1) + \Phi_V^{L+1} r_{2322}(L-1) + \Phi_V^L r_{2332}(L-1)]$$
$$-[\ p_{2,2}(L)I_m \quad (1-p_{2,2}(L))I_m\ ][\bar{H}\bar{\Phi}^L r_{3312}(L-1) + \Phi_V^{L+1} r_{3322}(L-1)$$
$$+\Phi_V^L r_{3332}(L-1)]\} \ , \quad r_{3312}(0) = 0$$

$$r_{3313}(L) = J_{31}(L,L)[\ 0 \quad p(L)I_m\ ]R_U(L)(\bar{\Phi}_V^T)^{-L} + r_{3313}(L-1)$$
$$-J_{31}(L,L)\{[\ p_{2,2}(L)I_m \quad (1-p_{2,2}(L))I_m\ ][\bar{H}\bar{\Phi}^L r_{1313}(L-1) + \Phi_V^{L+1} r_{1323}(L-1) + \Phi_V^L r_{1333}(L-1)]$$
$$+[\ p(L)I_m \quad (1-p(L))I_m\ ][\bar{H}\bar{\Phi}^L r_{2313}(L-1) + \Phi_V^{L+1} r_{2323}(L-1) + \Phi_V^L r_{2333}(L-1)]$$
$$-[\ p_{2,2}(L)I_m \quad (1-p_{2,2}(L))I_m\ ][\bar{H}\bar{\Phi}^L r_{3313}(L-1) + \Phi_V^{L+1} r_{3323}(L-1)$$
$$+\Phi_V^L r_{3333}(L-1)]\} \ , \quad r_{3313}(0) = 0$$

$$r_{3321}(L) = J_{32}(L,L)[\ 0 \quad p(L)I_m\ ]\bar{H}K_{\bar{x}}(L,L)(\bar{\Phi}^T)^{-L} + r_{3321}(L-1)$$
$$-J_{32}(L,L)\{[\ p_{2,2}(L)I_m \quad (1-p_{2,2}(L))I_m\ ][\bar{H}\bar{\Phi}^L r_{1311}(L-1) + \Phi_V^{L+1} r_{1321}(L-1) + \Phi_V^L r_{1331}(L-1)]$$
$$+[\ p(L)I_m \quad (1-p(L))I_m\ ][\bar{H}\bar{\Phi}^L r_{2311}(L-1) + \Phi_V^{L+1} r_{2321}(L-1) + \Phi_V^L r_{2331}(L-1)]$$
$$-[\ p_{2,2}(L)I_m \quad (1-p_{2,2}(L))I_m\ ][\bar{H}\bar{\Phi}^L r_{3311}(L-1) + \Phi_V^{L+1} r_{3321}(L-1)$$
$$+\Phi_V^L r_{3331}(L-1)]\} \ , \quad r_{3321}(0) = 0$$

$$r_{3322}(L) = J_{32}(L,L)[\ 0 \quad p(L)I_m\ ]\Phi_V K_V(L,L)(\bar{\Phi}_V^T)^{-L} + r_{3322}(L-1)$$
$$-J_{32}(L,L)\{[\ p_{2,2}(L)I_m \quad (1-p_{2,2}(L))I_m\ ][\bar{H}\bar{\Phi}^L r_{1312}(L-1) + \Phi_V^{L+1} r_{1322}(L-1) + \Phi_V^L r_{1332}(L-1)]$$
$$+[\ p(L)I_m \quad (1-p(L))I_m\ ][\bar{H}\bar{\Phi}^L r_{2312}(L-1) + \Phi_V^{L+1} r_{2322}(L-1) + \Phi_V^L r_{2332}(L-1)]$$
$$-[\ p_{2,2}(L)I_m \quad (1-p_{2,2}(L))I_m\ ][\bar{H}\bar{\Phi}^L r_{3312}(L-1) + \Phi_V^{L+1} r_{3322}(L-1)$$
$$+\Phi_V^L r_{3332}(L-1)]\} \ , \quad r_{3322}(0) = 0$$

$$r_{3323}(L) = J_{32}(L,L)[\ 0 \quad p(L)I_m\ ]R_U(L)(\bar{\Phi}_V^T)^{-L} + r_{3323}(L-1)$$
$$-J_{32}(L,L)\{[\ p_{2,2}(L)I_m \quad (1-p_{2,2}(L))I_m\ ][\bar{H}\bar{\Phi}^L r_{1313}(L-1) + \Phi_V^{L+1} r_{1323}(L-1) + \Phi_V^L r_{1333}(L-1)]$$
$$+[\ p(L)I_m \quad (1-p(L))I_m\ ][\bar{H}\bar{\Phi}^L r_{2313}(L-1) + \Phi_V^{L+1} r_{2323}(L-1) + \Phi_V^L r_{2333}(L-1)]$$
$$-[\ p_{2,2}(L)I_m \quad (1-p_{2,2}(L))I_m\ ][\bar{H}\bar{\Phi}^L r_{3313}(L-1) + \Phi_V^{L+1} r_{3323}(L-1)$$
$$+\Phi_V^L r_{3333}(L-1)]\} \ , \quad r_{3323}(0) = 0$$

$$r_{3331}(L) = J_{33}(L,L)[\ 0 \quad p(L)I_m\ ]\bar{H}K_{\bar{x}}(L,L)(\bar{\Phi}^T)^{-L} + r_{3331}(L-1)$$
$$-J_{33}(L,L)\{[\ p_{2,2}(L)I_m \quad (1-p_{2,2}(L))I_m\ ][\bar{H}\bar{\Phi}^L r_{1311}(L-1) + \Phi_V^{L+1} r_{1321}(L-1) + \Phi_V^L r_{1331}(L-1)]$$
$$+[\ p(L)I_m \quad (1-p(L))I_m\ ][\bar{H}\bar{\Phi}^L r_{2311}(L-1) + \Phi_V^{L+1} r_{2321}(L-1) + \Phi_V^L r_{2331}(L-1)]$$
$$-[\ p_{2,2}(L)I_m \quad (1-p_{2,2}(L))I_m\ ][\bar{H}\bar{\Phi}^L r_{3311}(L-1) + \Phi_V^{L+1} r_{3321}(L-1)$$
$$+\Phi_V^L r_{3331}(L-1)]\} \ , \quad r_{3331}(0) = 0$$

$$r_{3332}(L) = J_{33}(L,L)[\ 0 \quad p(L)I_m\ ]\Phi_V K_V(L,L)(\bar{\Phi}_V^T)^{-L} + r_{3332}(L-1)$$
$$-J_{33}(L,L)\{[\ p_{2,2}(L)I_m \quad (1-p_{2,2}(L))I_m\ ][\bar{H}\bar{\Phi}^L r_{1312}(L-1) + \Phi_V^{L+1} r_{1322}(L-1) + \Phi_V^L r_{1332}(L-1)]$$
$$+[\ p(L)I_m \quad (1-p(L))I_m\ ][\bar{H}\bar{\Phi}^L r_{2312}(L-1) + \Phi_V^{L+1} r_{2322}(L-1) + \Phi_V^L r_{2332}(L-1)]$$
$$-[\ p_{2,2}(L)I_m \quad (1-p_{2,2}(L))I_m\ ][\bar{H}\bar{\Phi}^L r_{3312}(L-1) + \Phi_V^{L+1} r_{3322}(L-1)$$
$$+\Phi_V^L r_{3332}(L-1)]\} \ , \quad r_{3332}(0) = 0$$

$$r_{3333}(L) = J_{33}(L,L)[\ 0 \quad p(L)I_m\ ]R_U(L)(\bar{\Phi}_V^T)^{-L} + r_{3333}(L-1)$$
$$-J_{33}(L,L)\{[\ p_{2,2}(L)I_m \quad (1-p_{2,2}(L))I_m\ ][\bar{H}\bar{\Phi}^L r_{1313}(L-1) + \Phi_V^{L+1} r_{1323}(L-1) + \Phi_V^L r_{1333}(L-1)]$$
$$+[\ p(L)I_m \quad (1-p(L))I_m\ ][\bar{H}\bar{\Phi}^L r_{2313}(L-1) + \Phi_V^{L+1} r_{2323}(L-1) + \Phi_V^L r_{2333}(L-1)]$$
$$-[\ p_{2,2}(L)I_m \quad (1-p_{2,2}(L))I_m\ ][\bar{H}\bar{\Phi}^L r_{3313}(L-1) + \Phi_V^{L+1} r_{3323}(L-1)$$
$$+\Phi_V^L r_{3333}(L-1)]\}\ , \quad r_{3333}(0) = 0$$

Now, the equation for $\bar{C}_{11}(L,L)$ is derived as follows.

$$\bar{C}_{11}(L,L) = C_{11}(L,L)(\bar{\Phi}^T)^L$$
$$= \sum_{i=1}^{L} h(L,i,L)[\ p(i)I_m \quad 0\ ]\bar{H}K_{\bar{x}}(i,i)(\bar{\Phi}^T)^{-i}(\bar{\Phi}^T)^L$$
$$= \bar{\Phi}^L \sum_{i=1}^{L} \bar{J}(i,L)[\ p(i)I_m \quad 0\ ]\bar{H}K_{\bar{x}}(i,i)(\bar{\Phi}^T)^{-i}(\bar{\Phi}^T)^L$$
$$= \bar{\Phi}^L \bar{J}(L,L)[\ p(L)I_m \quad 0\ ]\bar{H}K_{\bar{x}}(L,L) + \bar{\Phi}^L \sum_{i=1}^{L-1} \bar{J}(i,L)[\ p(i)I_m \quad 0\ ]\bar{H}K_{\bar{x}}(i,i)(\bar{\Phi}^T)^{-i}(\bar{\Phi}^T)^L$$
$$= \bar{\Phi}^L \bar{J}(L,L)[\ p(L)I_m \quad 0\ ]\bar{H}K_{\bar{x}}(L,L)$$
$$+\bar{\Phi}^L \sum_{i=1}^{L-1} \{\bar{J}(i,L-1) - \bar{J}(L,L)\{[\ p_{2,2}(L)I_m \quad (1-p_{2,2}(L))I_m\ ][\ \bar{H}\bar{\Phi}^L \quad \Phi_V^{L+1} \quad \Phi_V^L\ ]J_1(i,L-1)$$
$$+[\ p(L)I_m \quad (1-p(L))I_m 0\ ][\ \bar{H}\bar{\Phi}^L \quad \Phi_V^{L+1} \quad \Phi_V^L\ ]J_2(i,L-1)$$
$$-[\ p_{2,2}(L)I_m \quad (1-p_{2,2}(L))I_m\ ][\ \bar{H}\bar{\Phi}^L \quad \Phi_V^{L+1} \quad \Phi_V^L\ ]J_3(i,L-1)\}\}$$
$$[\ p(i)I_m \quad 0\ ]\bar{H}K_{\bar{x}}(i,i)(\bar{\Phi}^T)^{-i}(\bar{\Phi}^T)^L$$
$$= h(L,L,L)[\ p(L)I_m \quad 0\ ]\bar{H}K_{\bar{x}}(L,L)$$
$$+\bar{\Phi}^L \sum_{i=1}^{L-1} \{\bar{J}(i,L-1) - \bar{J}(L,L)\{[\ p_{2,2}(L)I_m \quad (1-p_{2,2}(L))I_m\ ][\ \bar{H}\bar{\Phi}^L \quad \Phi_V^{L+1} \quad \Phi_V^L\ ]\begin{bmatrix} J_{11}(i,L-1) \\ J_{12}(i,L-1) \\ J_{13}(i,L-1) \end{bmatrix}$$
$$+[\ p(L)I_m \quad (1-p(L))I_m 0\ ][\ \bar{H}\bar{\Phi}^L \quad \Phi_V^{L+1} \quad \Phi_V^L\ ]\begin{bmatrix} J_{21}(i,L-1) \\ J_{22}(i,L-1) \\ J_{23}(i,L-1) \end{bmatrix}$$
$$-[\ p_{2,2}(L)I_m \quad (1-p_{2,2}(L))I_m\ ][\ \bar{H}\bar{\Phi}^L \quad \Phi_V^{L+1} \quad \Phi_V^L\ ]\begin{bmatrix} J_{31}(i,L-1) \\ J_{32}(i,L-1) \\ J_{33}(i,L-1) \end{bmatrix}\}\}$$
$$[\ p(i)I_m \quad 0\ ]\bar{H}K_{\bar{x}}(i,i)(\bar{\Phi}^T)^{-i}(\bar{\Phi}^T)^L$$
$$= h(L,L,L)[\ p(L)I_m \quad 0\ ]\bar{H}K_{\bar{x}}(L,L)$$
$$+\bar{\Phi}^L B_{11}(L-1)(\bar{\Phi}^T)^L - \bar{\Phi}^L \bar{J}(L,L)\{[\ p_{2,2}(L)I_m \quad (1-p_{2,2}(L))I_m\ ][\bar{H}\bar{\Phi}^L r_{1111}(L-1)(\bar{\Phi}^T)^L$$
$$+\Phi_V^{L+1} r_{1121}(L-1)(\bar{\Phi}^T)^L + \Phi_V^L r_{1131}(L-1)(\bar{\Phi}^T)^L] +$$
$$+[\ p(L)I_m \quad (1-p(L))I_m 0\ ][\bar{H}\bar{\Phi}^L r_{2111}(L-1)(\bar{\Phi}^T)^L + \Phi_V^{L+1} r_{2121}(L-1)(\bar{\Phi}^T)^L$$
$$+\Phi_V^L r_{2131}(L-1)(\bar{\Phi}^T)^L]$$
$$-[\ p_{2,2}(L)I_m \quad (1-p_{2,2}(L))I_m\ ][\bar{H}\bar{\Phi}^L r_{3111}(L-1)(\bar{\Phi}^T)^L + \Phi_V^{L+1} r_{3121}(L-1)(\bar{\Phi}^T)^L$$
$$+\Phi_V^L r_{3131}(L-1)(\bar{\Phi}^T)^L]\}$$

$$= h(L,L,L)[\ p(L)I_m \quad 0\ ]\bar{H}K_{\tilde{x}}(L,L)$$
$$+\bar{\Phi}\tilde{B}_{11}(L-1)\bar{\Phi}^T - h(L,L,L)\{[\ p_{2,2}(L)I_m \quad (1-p_{2,2}(L))I_m\ ][\bar{H}\bar{\Phi}S_{1111}(L-1)\bar{\Phi}^T$$
$$+\Phi_V^2 S_{1121}(L-1)\bar{\Phi}^T + \Phi_V S_{1131}(L-1)\bar{\Phi}^T] + [\ p(L)I_m \quad (1-p(L))I_m\ ][\bar{H}\bar{\Phi}S_{2111}(L-1)\bar{\Phi}^T$$
$$+\Phi_V^2 S_{2121}(L-1)\bar{\Phi}^T + \Phi_V S_{2131}(L-1)\bar{\Phi}^T] - [\ p_{2,2}(L)I_m \quad (1-p_{2,2}(L))I_m\ ][\bar{H}\bar{\Phi}S_{3111}(L-1)\bar{\Phi}^T$$
$$+\Phi_V^2 S_{3121}(L-1)\bar{\Phi}^T + \Phi_V S_{3131}(L-1)\bar{\Phi}^T]\}$$

Similarly, we obtain the following equations.

$$\bar{C}_{12}(L,L) = h(L,L,L)[\ p(L)I_m \quad 0\ ]\bar{H}K_{\tilde{x}}(L,L)$$
$$+\bar{\Phi}\tilde{B}_{12}(L-1)\bar{\Phi}_V^T - h(L,L,L)\{[\ p_{2,2}(L)I_m \quad (1-p_{2,2}(L))I_m\ ][\bar{H}\bar{\Phi}S_{11112}(L-1)\Phi_V^T$$
$$+\Phi_V^2 S_{1122}(L-1)\Phi_V^T + \Phi_V S_{1132}(L-1)\Phi_V^T] + [\ p(L)I_m \quad (1-p(L))I_m 0\ ][\bar{H}\bar{\Phi}S_{2112}(L-1)\Phi_V^T$$
$$+\Phi_V^2 S_{2122}(L-1)\Phi_V^T + \Phi_V S_{2132}(L-1)\Phi_V^T] - [\ p_{2,2}(L)I_m \quad (1-p_{2,2}(L))I_m\ ][\bar{H}\bar{\Phi}S_{3112}(L-1)\Phi_V^T$$
$$+\Phi_V^2 S_{3122}(L-1)\Phi_V^T + \Phi_V S_{3132}(L-1)\Phi_V^T]\}$$

$$\bar{C}_{13}(L,L) = h(L,L,L)[\ p(L)I_m \quad 0\ ]R_U(L)$$
$$+\bar{\Phi}\tilde{B}_{13}(L-1)\bar{\Phi}_V^T - h(L,L,L)\{[\ p_{2,2}(L)I_m \quad (1-p_{2,2}(L))I_m\ ][\bar{H}\bar{\Phi}S_{11113}(L-1)\Phi_V^T$$
$$+\Phi_V^2 S_{1123}(L-1)\Phi_V^T + \Phi_V S_{1133}(L-1)\Phi_V^T] + [\ p(L)I_m \quad (1-p(L))I_m 0\ ][\bar{H}\bar{\Phi}S_{2113}(L-1)\Phi_V^T$$
$$+\Phi_V^2 S_{2123}(L-1)\Phi_V^T + \Phi_V S_{2133}(L-1)\Phi_V^T] - [\ p_{2,2}(L)I_m \quad (1-p_{2,2}(L))I_m\ ][\bar{H}\bar{\Phi}S_{3113}(L-1)\Phi_V^T$$
$$+\Phi_V^2 S_{3123}(L-1)\Phi_V^T + \Phi_V S_{3133}(L-1)\Phi_V^T]\}$$

$$\bar{C}_{21}(L,L) = h(L,L,L)[\ 0 \quad I_m\ ]\bar{H}K_{\tilde{x}}(L,L)$$
$$+\bar{\Phi}\tilde{B}_{21}(L-1)\bar{\Phi}^T - h(L,L,L)\{[\ p_{2,2}(L)I_m \quad (1-p_{2,2}(L))I_m\ ][\bar{H}\bar{\Phi}S_{1211}(L-1)\bar{\Phi}^T$$
$$+\Phi_V^2 S_{1221}(L-1)\bar{\Phi}^T + \Phi_V S_{1231}(L-1)\bar{\Phi}^T] + [\ p(L)I_m \quad (1-p(L))I_m\ ][\bar{H}\bar{\Phi}S_{2211}(L-1)\bar{\Phi}^T$$
$$+\Phi_V^2 S_{2221}(L-1)\bar{\Phi}^T + \Phi_V S_{2231}(L-1)\bar{\Phi}^T] - [\ p_{2,2}(L)I_m \quad (1-p_{2,2}(L))I_m\ ][\bar{H}\bar{\Phi}S_{3211}(L-1)\bar{\Phi}^T$$
$$+\Phi_V^2 S_{3221}(L-1)\bar{\Phi}^T + \Phi_V S_{3231}(L-1)\bar{\Phi}^T]\}$$

$$\bar{C}_{22}(L,L) = h(L,L,L)[\ 0 \quad I_m\ ]\bar{H}K_{\tilde{x}}(L,L)$$
$$+\bar{\Phi}\tilde{B}_{22}(L-1)\Phi_V^T - h(L,L,L)\{[\ p_{2,2}(L)I_m \quad (1-p_{2,2}(L))I_m\ ][\bar{H}\bar{\Phi}S_{1212}(L-1)\Phi_V^T$$
$$+\Phi_V^2 S_{1222}(L-1)\Phi_V^T + \Phi_V S_{1232}(L-1)\Phi_V^T] + [\ p(L)I_m \quad (1-p(L))I_m\ ][\bar{H}\bar{\Phi}S_{2212}(L-1)\Phi_V^T$$
$$+\Phi_V^2 S_{2222}(L-1)\Phi_V^T + \Phi_V S_{2232}(L-1)\Phi_V^T] - [\ p_{2,2}(L)I_m \quad (1-p_{2,2}(L))I_m\ ][\bar{H}\bar{\Phi}S_{3212}(L-1)\Phi_V^T$$
$$+\Phi_V^2 S_{3222}(L-1)\Phi_V^T + \Phi_V S_{3232}(L-1)\Phi_V^T]\}$$

$$\bar{C}_{23}(L,L) = h(L,L,L)[\ 0 \quad I_m\ ]R_U(L)$$
$$+\bar{\Phi}\tilde{B}_{23}(L-1)\Phi_V^T - h(L,L,L)\{[\ p_{2,2}(L)I_m \quad (1-p_{2,2}(L))I_m\ ][\bar{H}\bar{\Phi}S_{1213}(L-1)\Phi_V^T$$
$$+\Phi_V^2 S_{1223}(L-1)\Phi_V^T + \Phi_V S_{1233}(L-1)\Phi_V^T] + [\ p(L)I_m \quad (1-p(L))I_m 0\ ][\bar{H}\bar{\Phi}S_{2213}(L-1)\Phi_V^T$$
$$+\Phi_V^2 S_{2223}(L-1)\Phi_V^T + \Phi_V S_{2233}(L-1)\Phi_V^T] - [\ p_{2,2}(L)I_m \quad (1-p_{2,2}(L))I_m\ ][\bar{H}\bar{\Phi}S_{3213}(L-1)\Phi_V^T$$
$$+\Phi_V^2 S_{3223}(L-1)\Phi_V^T + \Phi_V S_{3233}(L-1)\Phi_V^T]\}$$

$$\bar{C}_{31}(L,L) = h(L,L,L)[\ 0 \quad p(L)I_m\ ]\bar{H}K_{\tilde{x}}(L,L)$$
$$+\bar{\Phi}\tilde{B}_{21}(L-1)\bar{\Phi}^T - h(L,L,L)\{[\ p_{2,2}(L)I_m \quad (1-p_{2,2}(L))I_m\ ][\bar{H}\bar{\Phi}S_{1311}(L-1)\bar{\Phi}^T$$
$$+\Phi_V^2 S_{1321}(L-1)\bar{\Phi}^T + \Phi_V S_{1331}(L-1)\bar{\Phi}^T] + [\ p(L)I_m \quad (1-p(L))I_m\ ][\bar{H}\bar{\Phi}S_{2311}(L-1)\bar{\Phi}^T$$
$$+\Phi_V^2 S_{2321}(L-1)\bar{\Phi}^T + \Phi_V S_{2331}(L-1)\bar{\Phi}^T] - [\ p_{2,2}(L)I_m \quad (1-p_{2,2}(L))I_m\ ][\bar{H}\bar{\Phi}S_{3311}(L-1)\bar{\Phi}^T$$
$$+\Phi_V^2 S_{3321}(L-1)\bar{\Phi}^T + \Phi_V S_{3331}(L-1)\bar{\Phi}^T]\}$$

$$\bar{C}_{32}(L,L) = h(L,L,L)[\ 0 \quad p(L)I_m\ ]\bar{H}K_{\bar{x}}(L,L)$$
$$+\bar{\Phi}\tilde{B}_{32}(L-1)\Phi_V^T - h(L,L,L)\{[\ p_{2,2}(L)I_m \quad (1-p_{2,2}(L))I_m\ ][\bar{H}\bar{\Phi}S_{1312}(L-1)\Phi_V^T$$
$$+\Phi_V^2 S_{1322}(L-1)\Phi_V^T + \Phi_V S_{1332}(L-1)\Phi_V^T] + [\ p(L)I_m \quad (1-p(L))I_m\ ][\bar{H}\bar{\Phi}S_{2312}(L-1)\Phi_V^T$$
$$+\Phi_V^2 S_{2322}(L-1)\Phi_V^T + \Phi_V S_{2332}(L-1)\Phi_V^T] - [\ p_{2,2}(L)I_m \quad (1-p_{2,2}(L))I_m\ ][\bar{H}\bar{\Phi}S_{3312}(L-1)\Phi_V^T$$
$$+\Phi_V^2 S_{3322}(L-1)\Phi_V^T + \Phi_V S_{3332}(L-1)\Phi_V^T]\}$$

$$\bar{C}_{33}(L,L) = h(L,L,L)[\ 0 \quad p(L)I_m\ ]R_U(L)$$
$$+\bar{\Phi}\tilde{B}_{33}(L-1)\Phi_V^T - h(L,L,L)\{[\ p_{2,2}(L)I_m \quad (1-p_{2,2}(L))I_m\ ][\bar{H}\bar{\Phi}S_{1313}(L-1)\Phi_V^T$$
$$+\Phi_V^2 S_{1323}(L-1)\Phi_V^T + \Phi_V S_{1333}(L-1)\Phi_V^T] + [\ p(L)I_m \quad (1-p(L))I_m\ ][\bar{H}\bar{\Phi}S_{2313}(L-1)\Phi_V^T$$
$$+\Phi_V^2 S_{2323}(L-1)\Phi_V^T + \Phi_V S_{2333}(L-1)\Phi_V^T] - [\ p_{2,2}(L)I_m \quad (1-p_{2,2}(L))I_m\ ][\bar{H}\bar{\Phi}S_{3313}(L-1)\Phi_V^T$$
$$+\Phi_V^2 S_{3323}(L-1)\Phi_V^T + \Phi_V S_{3333}(L-1)\Phi_V^T]\}$$

Now, it is seen that $B_{11}(L)$ is rewritten as

$$B_{11}(L) = \bar{J}(L,L)[\ p(L)I_m \quad 0\ ]\bar{H}K_{\bar{x}}(L,L)(\bar{\Phi}^T)^{-L} + \sum_{i=1}^{L-1}\bar{J}(i,L)[\ p(i)I_m \quad 0\ ]\bar{H}K_{\bar{x}}(i,i)(\bar{\Phi}^T)^{-i}$$
$$= \bar{J}(L,L)[\ p(L)I_m \quad 0\ ]\bar{H}K_{\bar{x}}(L,L)(\bar{\Phi}^T)^{-L}$$
$$+\sum_{i=1}^{L-1}\{\bar{J}(s,L-1) - \bar{J}(L,L)\{[\ p_{2,2}(L)I_m \quad (1-p_{2,2}(L))I_m\ ][\ \bar{H}\bar{\Phi}^L \quad \Phi_V^{L+1} \quad \Phi_V^L\ ]\begin{bmatrix} J_{11}(i,L-1) \\ J_{12}(i,L-1) \\ J_{13}(i,L-1) \end{bmatrix}$$
$$+[\ p(L)I_m \quad (1-p(L))I_m 0\ ][\ \bar{H}\bar{\Phi}^L \quad \Phi_V^{L+1} \quad \Phi_V^L\ ]\begin{bmatrix} J_{21}(i,L-1) \\ J_{22}(i,L-1) \\ J_{23}(i,L-1) \end{bmatrix}$$
$$-[\ p_{2,2}(L)I_m \quad (1-p_{2,2}(L))I_m\ ][\ \bar{H}\bar{\Phi}^L \quad \Phi_V^{L+1} \quad \Phi_V^L\ ]\begin{bmatrix} J_{31}(i,L-1) \\ J_{32}(i,L-1) \\ J_{33}(i,L-1) \end{bmatrix}\}\}[\ p(i)I_m \quad 0\ ]\bar{H}K_{\bar{x}}(i,i)(\bar{\Phi}^T)^{-i}$$

$$= \bar{J}(L,L)[\ p(L)I_m \quad 0\ ]\bar{H}K_{\bar{x}}(L,L)(\bar{\Phi}^T)^{-L} + B_{11}(L-1)$$
$$-\bar{J}(L,L)\{[\ p_{2,2}(L)I_m \quad (1-p_{2,2}(L))I_m\ ][\bar{H}\bar{\Phi}^L r_{1111}(L-1) + \Phi_V^{L+1}r_{1121}(L-1) + \Phi_V^L r_{1131}(L-1)]$$
$$+[\ p(L)I_m \quad (1-p(L))I_m 0\ ][\bar{H}\bar{\Phi}^L r_{2111}(L-1) + \Phi_V^{L+1}r_{2121}(L-1) + \Phi_V^L r_{2131}(L-1)]$$
$$-[\ p_{2,2}(L)I_m \quad (1-p_{2,2}(L))I_m\ ][\bar{H}\bar{\Phi}^L r_{3111}(L-1) + \Phi_V^{L+1}r_{3121}(L-1) + \Phi_V^L r_{3131}(L-1)]\}.$$

From $\tilde{B}_{11}(L) = \bar{\Phi}^L B_{11}(L)(\bar{\Phi}^T)^L$, $\tilde{B}_{11}(L)$ is updated by

$$\tilde{B}_{11}(L) = \bar{G}(L,L)[\ p(L)I_m \quad 0\ ]\bar{H}K_{\bar{x}}(L,L) + \bar{\Phi}B_{11}(L-1)\bar{\Phi}^T$$
$$-\bar{G}(L,L)\{[\ p_{2,2}(L)I_m \quad (1-p_{2,2}(L))I_m\ ][\bar{H}\bar{\Phi}S_{1111}(L-1)\bar{\Phi}^T + \Phi_V^2 S_{1121}(L-1)\bar{\Phi}^T$$
$$+\Phi_V S_{1131}(L-1)\bar{\Phi}^T]$$
$$+[\ p(L)I_m \quad (1-p(L))I_m 0\ ][\bar{H}\bar{\Phi}S_{2111}(L-1)\bar{\Phi}^T + \Phi_V^2 S_{2121}(L-1)\bar{\Phi}^T + \Phi_V S_{2131}(L-1)\bar{\Phi}^T]$$
$$-[\ p_{2,2}(L)I_m \quad (1-p_{2,2}(L))I_m\ ][\bar{H}\bar{\Phi}S_{3111}(L-1)\bar{\Phi}^T + \Phi_V^2 S_{3121}(L-1)\bar{\Phi}^T + \Phi_V S_{3131}(L-1)\bar{\Phi}^T]\}.$$

The equations updating $\tilde{B}_{12}(L)$, $\tilde{B}_{13}(L)$, $\tilde{B}_{21}(L)$, $\tilde{B}_{22}(L)$, $\tilde{B}_{23}(L)$, $\tilde{B}_{31}(L)$, $\tilde{B}_{32}(L)$ and $\tilde{B}_{33}(L)$ are given by (3.127)-(3-134).

Subtracting $C_{11}(k,L-1)$ from $C_{11}(k,L)$, we obtain

$$
\begin{aligned}
&C_{11}(k,L)-C_{11}(k,L-1)=h(k,L,L)[\ p(L)I_m \quad 0\ ]\bar{H}K_{\tilde{x}}(L,L)\bar{\Phi}^{-L}\\
&-h(k,L,L)\{[\ p_{2,2}(L)I_m \quad (1-p_{2,2}(L))I_m\ ][\bar{H}\bar{\Phi}^{L}r_{1111}(L-1)+\Phi_V^{L+1}r_{1121}(L-1)+\Phi_V^{L}r_{1131}(L-1)]\\
&+[\ p(L)I_m \quad (1-p(L))I_m\ ][\bar{H}\bar{\Phi}^{L}r_{2111}(L-1)+\Phi_V^{L+1}r_{2121}(L-1)+\Phi_V^{L}r_{2131}(L-1)]\\
&-[\ p_{2,2}(L)I_m \quad (1-p_{2,2}(L))I_m\ ][\bar{H}\bar{\Phi}^{L}r_{3111}(L-1)+\Phi_V^{L+1}r_{3121}(L-1)+\Phi_V^{L}r_{3131}(L-1)]\}.
\end{aligned}
$$

Similarly, the following recursive equations are derived.

$$
\begin{aligned}
&C_{12}(k,L)-C_{12}(k,L-1)=h(k,L,L)[\ p(L)I_m \quad 0\ ]\Phi_V K_V(L,L)(\Phi_V^{T})^{-L}\\
&-h(k,L,L)\{[\ p_{2,2}(L)I_m \quad (1-p_{2,2}(L))I_m\ ][\bar{H}\bar{\Phi}^{L}r_{1112}(L-1)+\Phi_V^{L+1}r_{1122}(L-1)+\Phi_V^{L}r_{1132}(L-1)]\\
&+[\ p(L)I_m \quad (1-p(L))I_m\ ][\bar{H}\bar{\Phi}^{L}r_{2111}(L-1)+\Phi_V^{L+1}r_{2121}(L-1)+\Phi_V^{L}r_{2131}(L-1)]\\
&-[\ p_{2,2}(L)I_m \quad (1-p_{2,2}(L))I_m\ ][\bar{H}\bar{\Phi}^{L}r_{3111}(L-1)+\Phi_V^{L+1}r_{3121}(L-1)+\Phi_V^{L}r_{3131}(L-1)]\}
\end{aligned}
$$

$$
\begin{aligned}
&C_{13}(k,L)-C_{13}(k,L-1)=h(k,L,L)[\ p(L)I_m \quad 0\ ]R_U(L)(\Phi_V^{T})^{-L}\\
&-h(k,L,L)\{[\ p_{2,2}(L)I_m \quad (1-p_{2,2}(L))I_m\ ][\bar{H}\bar{\Phi}^{L}r_{1113}(L-1)+\Phi_V^{L+1}r_{1123}(L-1)+\Phi_V^{L}r_{1133}(L-1)]\\
&+[\ p(L)I_m \quad (1-p(L))I_m\ ][\bar{H}\bar{\Phi}^{L}r_{2113}(L-1)+\Phi_V^{L+1}r_{2123}(L-1)+\Phi_V^{L}r_{2133}(L-1)]\\
&-[\ p_{2,2}(L)I_m \quad (1-p_{2,2}(L))I_m\ ][\bar{H}\bar{\Phi}^{L}r_{3113}(L-1)+\Phi_V^{L+1}r_{3123}(L-1)+\Phi_V^{L}r_{3133}(L-1)]\}.
\end{aligned}
$$

$$
\begin{aligned}
&C_{21}(k,L)-C_{21}(k,L-1)=h(k,L,L)[\ 0 \quad I_m\ ]\bar{H}K_{\tilde{x}}(L,L)\bar{\Phi}^{-L}\\
&-h(k,L,L)\{[\ p_{2,2}(L)I_m \quad (1-p_{2,2}(L))I_m\ ][\bar{H}\bar{\Phi}^{L}r_{1211}(L-1)+\Phi_V^{L+1}r_{1221}(L-1)+\Phi_V^{L}r_{1231}(L-1)]\\
&+[\ p(L)I_m \quad (1-p(L))I_m\ ][\bar{H}\bar{\Phi}^{L}r_{2211}(L-1)+\Phi_V^{L+1}r_{2221}(L-1)+\Phi_V^{L}r_{2231}(L-1)]\\
&-[\ p_{2,2}(L)I_m \quad (1-p_{2,2}(L))I_m\ ][\bar{H}\bar{\Phi}^{L}r_{3211}(L-1)+\Phi_V^{L+1}r_{3221}(L-1)+\Phi_V^{L}r_{3231}(L-1)]\}
\end{aligned}
$$

$$
\begin{aligned}
&C_{22}(k,L)-C_{22}(k,L-1)=h(k,L,L)[\ 0 \quad I_m\ ]\Phi_V K_V(L,L)(\bar{\Phi}_V^{T})\ ^{L}\\
&-h(k,L,L)\{[\ p_{2,2}(L)I_m \quad (1-p_{2,2}(L))I_m\ ][\bar{H}\bar{\Phi}^{L}r_{1212}(L-1)+\Phi_V^{L+1}r_{1222}(L-1)+\Phi_V^{L}r_{1232}(L-1)]\\
&+[\ p(L)I_m \quad (1-p(L))I_m\ ][\bar{H}\bar{\Phi}^{L}r_{2212}(L-1)+\Phi_V^{L+1}r_{2222}(L-1)+\Phi_V^{L}r_{2232}(L-1)]\\
&-[\ p_{2,2}(L)I_m \quad (1-p_{2,2}(L))I_m\ ][\bar{H}\bar{\Phi}^{L}r_{3212}(L-1)+\Phi_V^{L+1}r_{3222}(L-1)+\Phi_V^{L}r_{3232}(L-1)]\}
\end{aligned}
$$

$$
\begin{aligned}
&C_{23}(k,L)-C_{23}(k,L-1)=h(k,L,L)[\ 0 \quad I_m\ ]R_U(L)(\Phi_V^{T})^{-L}\\
&-h(k,L,L)\{[\ p_{2,2}(L)I_m \quad (1-p_{2,2}(L))I_m\ ][\bar{H}\bar{\Phi}^{L}r_{1213}(L-1)+\Phi_V^{L+1}r_{1223}(L-1)+\Phi_V^{L}r_{1233}(L-1)]\\
&+[\ p(L)I_m \quad (1-p(L))I_m\ ][\bar{H}\bar{\Phi}^{L}r_{2213}(L-1)+\Phi_V^{L+1}r_{2223}(L-1)+\Phi_V^{L}r_{2233}(L-1)]\\
&-[\ p_{2,2}(L)I_m \quad (1-p_{2,2}(L))I_m\ ][\bar{H}\bar{\Phi}^{L}r_{3213}(L-1)+\Phi_V^{L+1}r_{3223}(L-1)+\Phi_V^{L}r_{3233}(L-1)]\}
\end{aligned}
$$

$$
\begin{aligned}
&C_{31}(k,L)-C_{31}(k,L-1)=h(k,L,L)[\ 0 \quad p(L)I_m\ ]\bar{H}K_{\tilde{x}}(L,L)\bar{\Phi}^{-L}\\
&-h(k,L,L)\{[\ p_{2,2}(L)I_m \quad (1-p_{2,2}(L))I_m\ ][\bar{H}\bar{\Phi}^{L}r_{1311}(L-1)+\Phi_V^{L+1}r_{1321}(L-1)+\Phi_V^{L}r_{1331}(L-1)]\\
&+[\ p(L)I_m \quad (1-p(L))I_m\ ][\bar{H}\bar{\Phi}^{L}r_{2311}(L-1)+\Phi_V^{L+1}r_{2321}(L-1)+\Phi_V^{L}r_{2331}(L-1)]\\
&-[\ p_{2,2}(L)I_m \quad (1-p_{2,2}(L))I_m\ ][\bar{H}\bar{\Phi}^{L}r_{3311}(L-1)+\Phi_V^{L+1}r_{3321}(L-1)+\Phi_V^{L}r_{3331}(L-1)]\}
\end{aligned}
$$

$$
\begin{aligned}
&C_{32}(k,L)-C_{32}(k,L-1)=h(k,L,L)[\ 0 \quad p(L)I_m\ ]\Phi_V K_V(L,L)(\Phi_V^{T})\ ^{L}\\
&-h(k,L,L)\{[\ p_{2,2}(L)I_m \quad (1-p_{2,2}(L))I_m\ ][\bar{H}\bar{\Phi}^{L}r_{1312}(L-1)+\Phi_V^{L+1}r_{1322}(L-1)+\Phi_V^{L}r_{1332}(L-1)]\\
&+[\ p(L)I_m \quad (1-p(L))I_m\ ][\bar{H}\bar{\Phi}^{L}r_{2312}(L-1)+\Phi_V^{L+1}r_{2322}(L-1)+\Phi_V^{L}r_{2332}(L-1)]\\
&-[\ p_{2,2}(L)I_m \quad (1-p_{2,2}(L))I_m\ ][\bar{H}\bar{\Phi}^{L}r_{3312}(L-1)+\Phi_V^{L+1}r_{3322}(L-1)+\Phi_V^{L}r_{3332}(L-1)]\}
\end{aligned}
$$

$$
\begin{aligned}
&C_{33}(k,L)-C_{33}(k,L-1)=h(k,L,L)[\ 0 \quad p(L)I_m\ ]R_U(L)(\Phi_V^{T})^{-L}\\
&-h(k,L,L)\{[\ p_{2,2}(L)I_m \quad (1-p_{2,2}(L))I_m\ ][\bar{H}\bar{\Phi}^{L}r_{1313}(L-1)+\Phi_V^{L+1}r_{1323}(L-1)+\Phi_V^{L}r_{1333}(L-1)]\\
&+[\ p(L)I_m \quad (1-p(L))I_m\ ][\bar{H}\bar{\Phi}^{L}r_{2313}(L-1)+\Phi_V^{L+1}r_{2323}(L-1)+\Phi_V^{L}r_{2333}(L-1)]\\
&-[\ p_{2,2}(L)I_m \quad (1-p_{2,2}(L))I_m\ ][\bar{H}\bar{\Phi}^{L}r_{3313}(L-1)+\Phi_V^{L+1}r_{3323}(L-1)+\Phi_V^{L}r_{3333}(L-1)]\}
\end{aligned}
$$

It is shown, from $\bar{C}_{11}(k,L) = C_{11}(k,L)(\bar{\Phi}^T)^L$, that $\bar{C}_{11}(k,L)$ is updated by

$$\begin{aligned}
&\bar{C}_{11}(k,L) - \bar{C}_{11}(k,L-1)\bar{\Phi}^T = h(k,L,L)[\ p(L)I_m \quad 0\ ]\bar{H}K_{\bar{x}}(L,L) \\
&-h(k,L,L)\{[\ p_{2,2}(L)I_m \quad (1-p_{2,2}(L))I_m\ ][\bar{H}\bar{\Phi}S_{1111}(L-1)\bar{\Phi}^T + \Phi_V^2 S_{1121}(L-1)\bar{\Phi}^T \\
&+\Phi_V S_{1131}(L-1)\bar{\Phi}^T] \\
&+[\ p(L)I_m \quad (1-p(L))I_m\ ][\bar{H}\bar{\Phi}S_{2111}(L-1)\bar{\Phi}^T + \Phi_V^2 S_{2121}(L-1)\bar{\Phi}^T + \Phi_V S_{2131}(L-1)\bar{\Phi}^T] \\
&-[\ p_{2,2}(L)I_m \quad (1-p_{2,2}(L))I_m\ ][\bar{H}\bar{\Phi}S_{3111}(L-1)\bar{\Phi}^T + \Phi_V^2 S_{3121}(L-1)\bar{\Phi}^T + \Phi_V S_{3131}(L-1)\bar{\Phi}^T]\}.
\end{aligned}$$

Similarly, $\bar{C}_{12}(k,L)$, $\bar{C}_{13}(k,L)$, $\bar{C}_{21}(k,L)$, $\bar{C}_{22}(k,L)$, $\bar{C}_{23}(k,L)$, $\bar{C}_{31}(k,L)$, $\bar{C}_{32}(k,L)$ and $\bar{C}_{33}(k,L)$ are updated by (3.14)-(3.21).

From (A-17), the following equation is derived.

$$\begin{aligned}
&\begin{bmatrix} O_{11}(L) \\ O_{12}(L) \\ O_{13}(L) \end{bmatrix} = \begin{bmatrix} O_{11}(L-1) \\ O_{12}(L-1) \\ O_{13}(L-1) \end{bmatrix} + \begin{bmatrix} J_{11}(L,L) \\ J_{12}(L,L) \\ J_{13}(L,L) \end{bmatrix} \{y(L) \\
&-[\ p_{2,2}(L)I_m \quad (1-p_{2,2}(L))I_m\ ][\ \bar{H}\bar{\Phi}^L \quad \Phi_V^{L+1} \quad \Phi_V^L\ ]\begin{bmatrix} O_{11}(L-1) \\ O_{12}(L-1) \\ O_{13}(L-1) \end{bmatrix} \\
&-[\ p(L)I_m \quad (1-p(L))I_m\ ][\ \bar{H}\bar{\Phi}^L \quad \Phi_V^{L+1} \quad \Phi_V^L\ ]\begin{bmatrix} O_{21}(L-1) \\ O_{22}(L-1) \\ O_{23}(L-1) \end{bmatrix} \\
&+[\ p_{2,2}(L)I_m \quad (1-p_{2,2}(L))I_m\ ][\ \bar{H}\bar{\Phi}^L \quad \Phi_V^{L+1} \quad \Phi_V^L\ ]\begin{bmatrix} O_{31}(L-1) \\ O_{32}(L-1) \\ O_{33}(L-1) \end{bmatrix}\}
\end{aligned}$$

$$\begin{aligned}
&= \begin{bmatrix} O_{11}(L-1) \\ O_{12}(L-1) \\ O_{13}(L-1) \end{bmatrix} + \begin{bmatrix} J_{11}(L,L) \\ J_{12}(L,L) \\ J_{13}(L,L) \end{bmatrix} \{y(L) \\
&-[\ p_{2,2}(L)I_m \quad (1-p_{2,2}(L))I_m\ ][\bar{H}\bar{\Phi}^L O_{11}(L-1) + \Phi_V^{L+1} O_{12}(L-1) + \Phi_V^L O_{13}(L-1)] \\
&-[\ p(L)I_m \quad (1-p(L))I_m\ ][\bar{H}\bar{\Phi}^L O_{21}(L-1) + \Phi_V^{L+1} O_{22}(L-1) + \Phi_V^L O_{23}(L-1)] \\
&+[\ p_{2,2}(L)I_m \quad (1-p_{2,2}(L))I_m\ ][\bar{H}\bar{\Phi}^L O_{31}(L-1) + \Phi_V^{L+1} O_{32}(L-1) + \Phi_V^L O_{33}(L-1)]\}
\end{aligned}$$

Introducing $\tilde{J}_{11}(L,L) = \bar{\Phi}^L J_{11}(L,L)$, from $\bar{O}_{11}(L) = \bar{\Phi}^L O_{11}(L)$, we obtain

$$\begin{aligned}
&\bar{O}_{11}(L) = \bar{\Phi}\bar{O}_{11}(L-1) + \bar{\Phi}^L \tilde{J}_{11}(L,L)\{y(L) \\
&-[\ p_{2,2}(L)I_m \quad (1-p_{2,2}(L))I_m\ ][\bar{H}\bar{\Phi}\bar{O}_{11}(L-1) + \Phi_V^2 \bar{O}_{12}(L-1) + \Phi_V \bar{O}_{13}(L-1)] \\
&-[\ p(L)I_m \quad (1-p(L))I_m\ ][\bar{H}\bar{\Phi}\bar{O}_{21}(L-1) + \Phi_V^2 \bar{O}_{22}(L-1) + \Phi_V \bar{O}_{23}(L-1)] \\
&+[\ p_{2,2}(L)I_m \quad (1-p_{2,2}(L))I_m\ ][\bar{H}\bar{\Phi}\bar{O}_{31}(L-1) + \Phi_V^2 \bar{O}_{32}(L-1) + \Phi_V \bar{O}_{33}(L-1)]\}.
\end{aligned}$$

Similarly, by introducing $\tilde{J}_{11}(L,L) = \bar{\Phi}^L J_{11}(L,L)$, $\tilde{J}_{12}(L,L) = \bar{\Phi}^L J_{12}(L,L)$, $\tilde{J}_{13}(L,L) = \bar{\Phi}^L J_{13}(L,L)$, $\tilde{J}_{21}(L,L) = \bar{\Phi}^L J_{21}(L,L)$, $\tilde{J}_{22}(L,L) = \bar{\Phi}^L J_{22}(L,L)$, $\tilde{J}_{23}(L,L) = \bar{\Phi}^L J_{23}(L,L)$, $\tilde{J}_{31}(L,L) = \bar{\Phi}^L J_{31}(L,L)$, $\tilde{J}_{32}(L,L) = \bar{\Phi}^L J_{32}(L,L)$ and $\tilde{J}_{33}(L,L) =$

$\bar{\Phi}^L J_{33}(L,L)$, the recursive equations for $\bar{O}_{12}(L)$, $\bar{O}_{13}(L)$, $\bar{O}_{21}(L)$, $\bar{O}_{22}(L)$, $\bar{O}_{23}(L)$, $\bar{O}_{31}(L)$, $\bar{O}_{32}(L)$ and $\bar{O}_{33}(L)$ are derived as (3-26)-(3-33).

From (A-16), the fixed-point smoothing estimate is updated by

$$\begin{aligned}&\hat{x}(k,L)=\hat{x}(k,L-1)+h(k,L,L)(y(L)-\bar{p}_{2,2}(L)(\bar{H}\bar{\Phi}\bar{O}_{11}(L-1)+\Phi_V^2\bar{O}_{12}(L-1)\\&+\Phi_V\bar{O}_{13}(L-1))-\bar{p}(L)(\bar{H}\bar{\Phi}\bar{O}_{21}(L-1)+\Phi_V^2\bar{O}_{22}(L-1)+\Phi_V\bar{O}_{23}(L-1))\\&+\bar{p}_{2,2}(L)(\bar{H}\bar{\Phi}\bar{O}_{31}(L-1)+\Phi_V^2\bar{O}_{32}(L-1)+\Phi_V\bar{O}_{33}(L-1))).\end{aligned}$$

From $S_{1111}(L)=\bar{\Phi}^L r_{1111}(L)(\bar{\Phi}^T)^L$ and the equation updating $r_{1111}(L)$ from $r_{1111}(L-1)$, the following equation updating $S_{1111}(L)$ from $S_{1111}(L-1)$ is obtained.

$$\begin{aligned}&S_{1111}(L)=\bar{\Phi}^L J_{11}(L,L)[\ p(L)I_m \quad 0\ ]\bar{H}K_{\bar{x}}(L,L)(\bar{\Phi}^T)^{-L}(\bar{\Phi}^T)^L+\bar{\Phi}^L r_{1111}(L-1)(\bar{\Phi}^T)^L\\&-\bar{\Phi}^L J_{11}(L,L)\{[\ p_{2,2}(L)I_m \quad (1-p_{2,2}(L))I_m\ ][\bar{H}\bar{\Phi}^L r_{1111}(L-1)+\Phi_V^{L+1}r_{1121}(L-1)\\&+\Phi_V^L r_{1131}(L-1)]\\&+[\ p(L)I_m \quad (1-p(L))I_m\ ][\bar{H}\bar{\Phi}^L r_{2111}(L-1)+\Phi_V^{L+1}r_{2121}(L-1)+\Phi_V^L r_{2131}(L-1)]\\&-[\ p_{2,2}(L)I_m \quad (1-p_{2,2}(L))I_m\ ][\bar{H}\bar{\Phi}^L r_{3111}(L-1)+\Phi_V^{L+1}r_{3121}(L-1)+\Phi_V^L r_{3131}(L-1)]\}(\bar{\Phi}^T)^L\\&=\tilde{J}_{11}(L,L)[\ p(L)I_m \quad 0\ ]\bar{H}K_{\bar{x}}(L,L)(\bar{\Phi}^T)^{-L}(\bar{\Phi}^T)^L+S_{1111}(L-1)\\&-\tilde{J}_{11}(L,L)\{[\ p_{2,2}(L)I_m \quad (1-p_{2,2}(L))I_m\ ][\bar{H}\bar{\Phi}S_{1111}(L-1)\bar{\Phi}^T+\Phi_V^2 S_{1121}(L-1)\bar{\Phi}^T\\&+\Phi_V S_{1131}(L-1)\bar{\Phi}^T]\\&+[\ p(L)I_m \quad (1-p(L))I_m\ ][\bar{H}\bar{\Phi}S_{2111}(L-1)\bar{\Phi}^T+\Phi_V^2 S_{2121}(L-1)\bar{\Phi}^T+\Phi_V S_{2131}(L-1)\bar{\Phi}^T]\\&-[\ p_{2,2}(L)I_m \quad (1-p_{2,2}(L))I_m\ ][\bar{H}\bar{\Phi}S_{3111}(L-1)\bar{\Phi}^T+\Phi_V^2 S_{3121}(L-1)\bar{\Phi}^T\\&+\Phi_V S_{3131}(L-1)\bar{\Phi}^T]\}\ , \quad S_{1111}(0)=0\end{aligned}$$

Similarly, the equations updating $S_{1112}(L)$, $S_{1113}(L)$, $S_{1121}(L)$, $S_{1122}(L)$, $S_{1123}(L)$, $S_{1131}(L)$, $S_{1132}(L)$, $S_{1133}(L)$, $S_{1211}(L)$, $S_{1212}(L)$, $S_{1213}(L)$, $S_{1221}(L)$, $S_{1222}(L)$, $S_{1223}(L)$, $S_{1231}(L)$, $S_{1232}(L)$, $S_{1233}(L)$, $S_{1311}(L)$, $S_{1312}(L)$, $S_{1313}(L)$, $S_{1321}(L)$, $S_{1322}(L)$, $S_{1323}(L)$, $S_{1331}(L)$, $S_{1332}(L)$, $S_{1333}(L)$, $S_{2111}(L)$, $S_{2112}(L)$, $S_{2113}(L)$, $S_{2121}(L)$, $S_{2122}(L)$, $S_{2123}(L)$, $S_{2131}(L)$, $S_{2132}(L)$, $S_{2133}(L)$,$S_{2211}(L)$, $S_{2212}(L)$, $S_{2213}(L)$, $S_{2221}(L)$, $S_{2222}(L)$, $S_{2223}(L)$, $S_{2231}(L)$, $S_{2232}(L)$, $S_{2233}(L)$, $S_{2311}(L)$, $S_{2312}(L)$, $S_{2313}(L)$, $S_{2321}(L)$, $S_{2322}(L)$, $S_{2323}(L)$, $S_{2331}(L)$, $S_{2332}(L)$, $S_{2333}(L)$, $S_{3111}(L)$, $S_{3112}(L)$, $S_{3113}(L)$, $S_{3121}(L)$, $S_{3122}(L)$, $S_{3123}(L)$, $S_{3131}(L)$, $S_{3132}(L)$, $S_{3133}(L)$, $S_{3211}(L)$, $S_{3212}(L)$, $S_{3213}(L)$, $S_{3221}(L)$, $S_{3222}(L)$, $S_{3223}(L)$, $S_{3231}(L)$, $S_{3232}(L)$, $S_{3233}(L)$, $S_{3311}(L)$, $S_{3312}(L)$, $S_{3313}(L)$, $S_{3321}(L)$, $S_{3322}(L)$, $S_{3323}(L)$, $S_{3331}(L)$, $S_{3332}(L)$, $S_{3333}(L)$ are derived as (3.44)-(3.123).

(Q.E.D.)

References

[1] N Nahi, Optimal recursive estimation with uncertain observations, *IEEE Trans. Information Theory*, IT-15 (1969) 457-462.

[2] W. NaNacara and E.E. Yaz, Recursive estimation for linear and nonlinear systems with uncertain observations, *Signal Processing*, 62 (1997) 215-228.

[3] A. Ray, L. W. Liou and J. H. Shen, State estimation using randomly delayed measurements, *Journal of Dynamic Systems, Measurement, and Control*, 115 (1993) 19-26.

[4] P. Wu, E. E. Yaz and K. J. Olejniczak, Harmonic estimation with random sensor delay, Proc. 36th *International Conference on Decision and Control* (1997) 1524-1525.

[5] E. Yaz and A. Ray, Linear unbiased state estimation under randomly varying bounded sensor delay, *Applied Mathematics Letters*, 11 (1998) 27-32.

[6] Y. I. Yaz, E. E. Yaz and M. J. Mohseni, LMI-based state estimation for some discrete-time stochastic models, *Proc. 1998 IEEE International Conference on Control Applications* (1998) 456-460.

[7] S. Nakamori, Recursive estimation technique of signal from output measurement data in linear discrete-time systems, *IEICE Trans. Fundamentals*, E-78-A (1995) 600-607.

[8] S. Nakamori, R. C. 'Aguila, A. H. Carazo and J. L. P'erez, New design of estimators using covariance information with uncertain observations in linear discrete-time systems, *Applied Mathematics and Computation*, 135 (2003) 429-441.

[9] H. L. Trees, Detection, *Estimation and Modulation Theory*, Part 1, Wiley, New York, 1968.

[10] S. Nakamori, R. C. 'Aguila, A. H. Carazo and J. L. P'erez, Estimation algorithm from delayed measurements with correlation between signal and noise using covariance information, *IEICE Trans. Fundamentals*, E-87-A (2004) 1219-1225.

[11] S. Nakamori, A. H. Carazo and J. L. P'erez, Least-squares linear smoothers from randomly delayed observations with correlation in the delay, *IEICE Trans Fundamentals*, E89-A (2006) 486-493.

[12] C. Han and H. Zhang, Optimal state estimation for discrete-time systems with random observation delays, *Acta Automatica Sinica*, 35 (2009) 1446-1451.

[13] C. Han and H. Zhang, Linear optimal filtering for discrete-time systems with random jump delays, *Signal Processing*, 89 (2009) 1121-1128.

[14] S. Nakamori, A. Hermoso-Carazo and J. Linares-P'erez, Design of RLS Wiener estimators from randomly delayed observations in linear discrete-time stochastic systems, *Applied Mathematics and Computation*, 217 (2010) 3801-3815.

[15] S. Nakamori, R. Caballero-'Aguila, A. Hermoso-Carazo and J. Linares-P'erez, Linear recursive discrete-time estimators using covariance information under uncertain observations, *Signal Processing*, 83 (2003) 1553-1559.

[16] N. Hadidi and S. Schwartz, Linear recursive state estimation under uncertain observations, *IEEE Trans. Automatic Control*, AC-24 (1979) 944-948.

[17] S. Nakamori, R. Caballero-'Aguila, A. Hermoso-Carazo and J. Linares-P'erez, RLS Wiener predictor with uncertain observations in linear discrete-time stochastic systems, *J. of Signal and Information Processing*, 2 (2011) 152-158.

[18] A. P. Sage and J.L. Melsa, *Estimation Theory with Applications to Communications and Control*, McGraw-Hill, New York, 1971.

Index